# AS/A-LEVEL YEAR 1

## STUDENT GUIDE

## EDEXCEL

# Geography

Globalisation

Shaping places

Cameron Dunn

**HODDER**
EDUCATION
AN HACHETTE UK COMPANY

Hodder Education, an Hachette UK company, Blenheim Court, George Street, Banbury, Oxfordshire OX16 5BH

*Orders*

Bookpoint Ltd, 130 Park Drive, Milton Park, Abingdon, Oxfordshire OX14 4SB

tel: 01235 827827

fax: 01235 400401

e-mail: education@bookpoint.co.uk

Lines are open 9.00 a.m.–5.00 p.m., Monday to Saturday, with a 24-hour message answering service. You can also order through the Hodder Education website: www.hoddereducation.co.uk

© Cameron Dunn 2017

ISBN 978-1-4718-6409-4

First printed 2017

Impression number 5 4 3 2 1

Year 2020 2019 2018 2017

Cover photo: Kevin Eaves/Fotolia

Typeset by Integra Software Services Pvt Ltd, Pondicherry, India

Printed in Slovenia

Hachette UK's policy is to use papers that are natural, renewable and recyclable products and made from wood grown in sustainable forests. The logging and manufacturing processes are expected to conform to the environmental regulations of the country of origin.

# Contents

# ■Getting the most from this book

## Exam tips

Advice on key points in the text to help you learn and recall content, avoid pitfalls, and polish your exam technique in order to boost your grade.

## Knowledge check

Rapid-fire questions throughout the Content Guidance section to check your understanding.

## Knowledge check answers

**1** Turn to the back of the book for the Knowledge check answers.

## Summaries

- Each core topic is rounded off by a bullet-list summary for quick-check reference of what you need to know.

**e** 6/6 marks awarded The answer to Part (a) is correct. Any answer indicating an income source other than farming would gain 1 mark. The student correctly identifies the values for the two years from Figure 2 for Part (b)(i) and shows these in their answer, as well as working out the difference between them correctly, scoring 2 marks. Mark scheme questions such as this have an 'acceptable' range of correct answers usually +/- 5% around the precise answer. Notice that in Part (b)(ii) the basic impact — loss of jobs — is followed by two further points related to this reason, so this answer scored 3 marks.

**(c)** Explain two criteria that can be used to judge the success of regeneration schemes. (4 marks)

**(d)** Explain the importance of re-imaging as part of the wider regeneration process. (6 marks)

**e** Part (c) is a point-marked question. Care needs to be taken to choose two different criteria, such as deprivation levels and demographic change. In addition 2+2 developed explanations are needed rather than a long explanation of only one criterion. Part (d) is a Level-marked question (see Levels mark scheme on page 76). Answers need to demonstrate an understanding of re-imaging, but also how this fits into a broader physical and economic regeneration strategy. An example could be used to illustrate this answer.

**Student answer**

**(c)** The Index of Multiple Deprivation could be used to judge the success of regeneration schemes because changes in the income or crime domains between two dates, such as 2010 and 2015, could show improving economic or social conditions as a result of regeneration. Population numbers from the census could be used as a growing population indicates a successful area attracting people and investment, especially if growth is among younger working age people aged 20–30.

**(d)** Regeneration involves not only rebuilding infrastructure and developing new residential and commercial buildings but also changing the image of a place. This re-imaging aims to change the external perceptions of a place, which changes people's view of the area and helps attract investment. Both in Liverpool and Salford Quays, regeneration has involved re-imaging to associate the areas with arts and culture, such as the Lowry and MediaCity in Salford Quays, to change the perception away from an image of a run-down deindustrialised area. In some cases the use of branding and logos is important for use on advertising. However, on its own a new image does not lead to successful regeneration because it cannot improve quality of life for existing residents. This requires improvements to the built environment, services and job opportunities.

## Commentary on sample student answers

Read the comments (preceded by the icon **e**) showing how many marks each answer would be awarded in the exam and exactly where marks are gained or lost.

## Exam-style questions

## Commentary on the questions

Tips on what you need to do to gain full marks, indicated by the icon **e**

## Sample student answers

Practise the questions, then look at the student answers that follow.

# ■ About this book

Much of the knowledge and understanding needed for AS and A-level geography builds on what you have learned for GCSE geography, but with an added focus on key geographical concepts and depth of knowledge and understanding of content. This guide offers advice for the effective revision of human geography, which all students need to complete.

The first part of the A-level Paper 2 tests your knowledge and application of aspects of human geography with a particular focus on **Globalisation** and **Shaping places**, choosing *one* of the options **Regenerating places** and **Diverse places**. The whole exam (including the other areas of study not covered here) lasts 2 hours and 15 minutes and makes up 30% of the A-level qualification. The same topics and approach make up 50% of the AS Paper 2, which lasts 1 hour and 45 minutes. More information on the external exam papers is given in the Question & Answer section at the back of this book.

To be successful in this unit you have to understand:

- the key ideas of the content
- the nature of the assessment material — by reviewing and practising sample structured questions
- how to achieve a high level of performance within the exams.

This guide has two sections:

**Content Guidance** — this summarises some of the key information that you need to know to be able to answer the examination questions with a high degree of accuracy and depth. In particular, the meaning of key terms is made clear and some attention is paid to providing details of case study material to help to meet the spatial context requirement within the specification.

**Questions & Answers** — this includes some sample questions similar in style to those you might expect in the exam. There are some sample student responses to these questions as well as detailed analysis, which will give further guidance in relation to what exam markers are looking for to award top marks.

The best way to use this book is to read through the relevant topic area first before practising the questions. Only refer to the answers and examiner comments after you have attempted the questions.

# Content Guidance

This section outlines the following areas of the AS Geography and A-level Geography specifications:

■ Globalisation
■ Shaping places
  – Regenerating places
  – Diverse places

Read through the topic area before attempting a question from the Question & Answers section.

# ■ Globalisation

## What are the causes of globalisation and why has it accelerated in recent decades?

■ The process of globalisation has created a 'shrinking world' where people and places are ever more closely connected.
■ Key decision makers including governments and international organisations have played important roles in globalisation.
■ Some groups, and places, have been more affected by globalisation than others.

## What is globalisation?

**Globalisation** is a process that has caused the world to become more connected over time. It is a process that began centuries ago and which continues today. In the last few decades, the process has speeded up.

The connections are best thought of as 'flows' that can be of:

■ Goods: the products and commodities we all buy in shops, many of which were grown or made in distant places.
■ Capital: flows of money between people, banks, businesses and governments.
■ Information: such as data transferred between businesses and people, often using the internet.
■ People: flows of migrants and tourists from one part of the world to another.

Globalisation can take several forms, all involving increasing **interdependence** and interconnectedness, which are summarised in Table 1.

> ### Exam tip
> Learn a definition of globalisation for the exam, as it can be a hard term to define 'off the top of your head'.

**Globalisation** is defined by the *Financial Times* as 'a process by which national and regional economies, societies, and cultures have become integrated through the global network of trade, communication, immigration and transportation'.

**Interdependence** means the success of one place depends on the success of other places. Economic problems in one country can quickly spread to its trading partners and affect people in distant places.

**Table 1** The many forms of globalisation

| | |
|---|---|
| **Economic globalisation** involves the growth of global transnational corporations (TNCs), which have a global presence and global brand image. It also involves the spreading of investment around the globe and rapid growth in world trade. | **Cultural globalisation** involves people increasingly eating similar food, wearing similar clothes, listening to similar music and sharing similar values, many of which are 'western' in origin, i.e. from North America and Europe. |
| **Political globalisation** takes the form of the dominance of western democracies in political and economic decision making. It also spreads the view that democratic, consumerist societies are the most 'successful'. | **Demographic globalisation** occurs as migration and tourism increase: populations are becoming ever more fluid and mixed. |
| As problems such as global warming become ever more pressing, **environmental globalisation** is growing — the realisation that global environmental threats require global solutions. | |

Back in the 1960s most UK citizens holidayed in the UK or perhaps Spain, their clothes were made in the UK and curry or pasta were considered exotic foods. Today you eat food from all over the world, think nothing of holidaying in Dubai or Florida and most of your clothes are made in China, Vietnam or Bangladesh.

Globalisation has become wider because even recently isolated places such as Sub-Saharan Africa are increasingly connected to the rest of the world through trade and tourism. It has become deeper because in the developed world our food, goods, media, music, friends and places we visit are increasingly global.

## Transport and trade developments

A key factor driving globalisation, and accelerating it, has been developments in transport technology. These have encouraged growth in **trade**, as transporting goods and people around the world has become cheaper over time.

Figure 1 shows how developments in transport have created a '**shrinking world**'. The speed and ease of moving around the world has reduced the friction of distance between places as well as dramatically lowering the cost of trade. The development of shipping containers pioneered in the 1960s is perhaps the most important development. These ubiquitous steel boxes transport most consumer goods by ship:

- Before containers, cargo was loaded in crates or sacks, manually, into the holds of ships. Now containers are loaded and unloaded by crane, increasingly automatically.
- Containers are inter-modal, meaning they can be transported by ship, lorry or train.
- The world's fleet of 9500 container ships can carry up to 18,000 twenty-foot shipping containers each.

Container ships are so efficient that the transport costs of moving an iPhone or television from China to the UK are less than £1.

**Exam tip**

Make sure you recognise that globalisation is a process that has developed over centuries. It did not just start with the invention of the internet!

**Trade** means the exchanges of goods and services between people and companies, and is increasingly cross-border between countries rather than just within a country.

The '**shrinking world**' concept is the idea that the world in 2016 feels smaller than in 1916, because places are closer in terms of travel time and knowledge of distant places is widespread so they feel less 'exotic'.

| 1830s | | 1840s | | 1950s |
| --- | --- | --- | --- | --- |
| **Railways** | **Telegraph** | **Steam ships** | **Jet passenger aircraft** | **Containerisation** |
| Steam trains quickly replaced horse-drawn and canal transport | Electric telegraph was the first long-distance, instant communication technology | Replaced sailing ships and increased speed and cargo capacity dramatically | Reduced travel time for passengers to hours rather than days, replacing steam ships | Dramatically speeded up goods trade and reduced costs, making consumer goods cheaper |

**Figure 1** Developments in transport technology

## The communication revolution

The late twentieth and early twenty-first centuries have been dominated by developments in ICT (information and communication technology) and mobile technology. Little changed in terms of global communication between the adoption of the electric telegraph and the growth of landline telephones after 1900. However, developments have been very rapid since 1990:

- Mobile phones became widespread from the mid-1990s (the first text message was sent in 1992) and are now common even in many developing countries.
- Internet access became common from the mid-1990s, followed by fast broadband; now close to 50% of the world's population uses the internet.
- The global network of land-based and subsea fibre optic cables has allowed instant, global communication.
- Satellite-based television has meant popular channels are available worldwide, in many languages.

As mobile communications have grown, so the technology to use it has developed in terms of smartphones, tablets and smart watches. A huge number of social networking applications (Facebook 2006, Skype 2003, Instagram 2010, WhatsApp 2010) means that people can stay in touch and communicate as never before and experience **time-space compression**.

It is important to recognise that the internet and mobile communications revolution is not just important to individuals, but also to businesses because they can:

- keep in touch with all parts of their production, supply and sales network locally and globally
- transfer money and investments instantly
- instantly analyse data on sales, employees and orders from anywhere within their business.

A consequence of internet use is that many activities that were once 'face to face' or 'voice to voice' are now done without any person-to-person interaction. This includes banking — most people use electronic banking apps — booking hotels online and, increasingly, shopping.

**Knowledge check 1**

What types of transport can containers be used on?

**Time-space compression** is the idea that the cost, in terms of time or money, of communicating over distance has fallen rapidly, so the idea of someone being 'a long way away' is now largely irrelevant in terms of the ability to communicate with them.

**Knowledge check 2**

What technology carries internet data across oceans?

# International political and economic decision making

A number of different organisations have been important in promoting globalisation. It would be wrong to assume that globalisation 'has just happened'. Figure 2 shows the growth in world exports of goods (clothes, cars, food, electronics, etc.) from 1970 to 2014. Notice:

- increases in world trade in the late 1970s and mid-1990s
- the huge growth in export trade after 2002
- the sharp dip 2008–09: this was the global recession/global financial crisis
- the return to 'normal' export levels by 2011 but slow growth since.

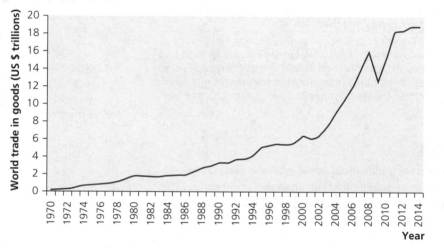

**Figure 2** Global exports of goods 1970–2014

Figure 2 illustrates the importance of trade to globalisation. A number of organisations have helped promote free trade and end 'protectionism'. In the past many countries protected their own industries and businesses by:

- demanding payment of taxes and tariffs on imported goods, so making them more expensive than home-produced goods
- using quotas to limit the volume of imports, protecting home producers from foreign competition
- banning foreign firms from operating in services like banking, retail and insurance
- restricting, or banning, foreign companies from investing in their country.

Protectionism reduces total trade volume, whereas free trade (no taxes, tariffs or quotas) increases it. Table 2 summarises the role of international economic and political organisations in promoting free trade policies and foreign direct investment (**FDI**).

> **Exam tip**
>
> Free trade and fair trade are often confused in the exam. Make sure you know what free trade means.

**Foreign Direct Investment (FDI)** is when a business from one country invests in another such as opening a chain of shops or building a factory.

**Table 2** The role of international organisations in free trade and FDI

| | |
|---|---|
| **World Trade Organisation (WTO)** | The international organisation that works to reduce trade barriers and create free trade. WTO was known as GATT (General Agreement on Tariffs and Trade) until 1995. A series of global agreements has gradually reduced trade barriers and increased free trade, although the latest round of talks began in Doha in 2001 and have not been agreed yet. |
| **International Monetary Fund (IMF)** | Since 1945 the IMF has worked to promote global economic and financial stability, and encourage more open economies. Part of this involves encouraging developing countries to accept FDI and open up their economies to free trade. The IMF has been criticised for promoting a 'western' model of economic development that works in the interests of developed countries and their TNCs. |
| **World Bank (WB)** | The World Bank's role since 1944 has been to lend money to the developing world to fund economic development and reduce poverty. It has helped developing countries develop deeper ties to the global economy but has been criticised for having policies that put economic development before social development. |

## National governments

Governments can choose to be part of a globalised world economy and most do. Famously, North Korea is a 'refuser' — sealing itself off from the world since 1953. Most governments actively seek global connections in the belief that trade promotes economic development and wealth. Governments can promote globalisation in a number of ways:

■ Joining **free trade blocs** such as the European Union (EU) and Association of South East Asian Nations (ASEAN), which make trade barrier-free between member states and in the case of the EU allows free movement of people between countries.

■ Opening up markets to competition: in many countries certain industries are protected or even operate as a monopoly such as national rail networks, postal services or electricity generation. Since 1980 there has been a move towards **free market liberalisation**, which has created competition in once restricted markets.

■ Privatisation: since the 1980s many governments have sold off industries they once owned (so-called 'nationalised industries'). In the UK the steel, car, electricity, gas and water industries were all state-owned but are now all privately owned. However, many governments still own big slices of industry even in EU countries such as France.

■ Grants and loans are often made to new businesses (called business start-ups) especially in areas that are seen to be globally important growth areas such as ICT development, pharmaceuticals or renewable energy. The UK Government's support for ICT start-ups in Tech City (sometimes called Silicon Roundabout) in the Old Street area of London falls into this category.

**Knowledge check 3**

Which global organisation has been mainly responsible for removing trade barriers between countries?

A **free trade bloc** is an agreement between a group of countries to remove all barriers to trade, e.g. import/export taxes, tariffs and quotas.

**Free market liberalisation** means ending monopoly provision of some services like telephones, broadband, gas and electricity, so you can choose your supplier based on quality and price.

## Special economic zones

In emerging countries there is now a long history of attempts by governments to promote particular regions as ideal locations for FDI. Beginning around 1980, countries such as China, India, Mexico and the Philippines began to create special economic zones (SEZs), free-trade zones (FTZs) or export processing zones (EPZs). China led the way in this area when in 1978 it decided on an Open Door Policy towards FDI and in 1980 created the Shenzhen Special Economic Zone. About 50 million people in more than 100 countries work in such locations.

SEZs and similar models are attractive to FDI for a number of reasons:
- They are tariff and quota free, allowing manufactured goods to be exported at no cost.
- Unions are usually banned, so workers cannot strike or complain.
- Infrastructure such as port facilities, roads, power and water connections are provided by the government, providing a subsidy for investors and lowering their costs.
- All profits made can be sent to the company HQ overseas.
- Taxes are usually very low, and often there is a tax-free period of up to 10 years after a business invests.
- Environmental regulations are usually limited.

SEZs have contributed hugely to 'made in China' as FDI has poured into that country in the last 30 years. Western consumers benefit from low-cost goods, but there are question marks about pay and working conditions in SEZs. Apple was subject to negative publicity in 2010 when working conditions in its supplier factories (owned by Foxconn) making iPhones and iPads came under scrutiny. In many Chinese SEZs wages are now high by global standards and countries like Vietnam are more competitive.

## The effects of globalisation

Not all places are globalised and their degree of globalisation varies. The KOF Index (Figure 3) measures the degree of globalisation of countries on an annual basis. It measures three aspects of globalisation:
- Economic globalisation measured by cross-border trade, investment and money flows.
- Social globalisation measured by international telephone calls, tourist flows, resident foreign population and access to foreign internet, TV, media and brands.
- Political globalisation measured by foreign embassies in a country, the number of international organisations the country is a member of and trade and other agreements with foreign countries.

**Exam tip**

Be prepared to discuss both the costs and benefits of special economic zones for the countries that use them.

**Knowledge check 4**

When did China first open the country up to foreign direct investment?

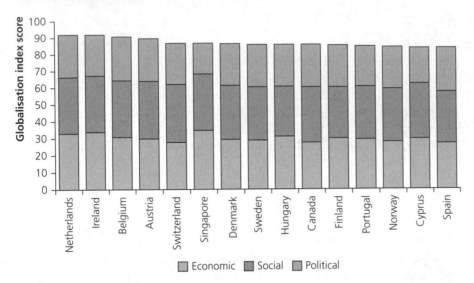

**Figure 3** The most globalised countries from the 2016 KOF index

Figure 3 shows that the most globalised countries tend to be European, relatively small and often involved in import/export trade. Many of the most globalised countries have culturally mixed populations and have many of their residents living abroad, as well as foreigners living in their country. The USA is less globalised (ranked 34 in 2016) than might be expected but this reflects the large parts of the interior of the USA that are not well connected to the rest of the world.

The AT Kearney Global Cities Index measures how economically successful cities are. In 2016 London, New York, Paris, Tokyo and Hong Kong were ranked as the most successful global cities, reflecting their global political importance as well as their role in the global financial system.

## The role of TNCs

One of the main drivers in globalisation has been **TNCs**. These major companies have a global 'reach'. TNCs are important creators of wealth. As Table 3 shows, the largest TNCs have turnovers (sales) equivalent in size to those of large countries. Some, such as Wal-Mart, employ enough people to populate an entire city. This gives them huge power. Investment decisions made by TNCs can be both a blessing and a curse.

**TNCs** are transnational corporations — companies that operate in more than one country.

**Table 3** The world's five largest TNCs compared to countries

| World's 5 largest companies, 2016 | Turnover in 2016 ($ billions) | Employees 2016 | Equivalent country by total GDP ($) 2016 |
|---|---|---|---|
| Wal-Mart Stores | 482 | 2,300,000 | Poland (472) |
| Royal Dutch Shell | 272 | 90,000 | Pakistan (269) |
| Exxon Mobil | 246 | 75,600 | Ireland (254) |
| Volkswagen | 236 | 610,000 | Chile (235) |
| Toyota | 236 | 346,000 | Finland (234) |

Much of China's rapid economic growth has been fuelled by western TNCs locating manufacturing plants in its SEZs, creating jobs and boosting exports, taking advantage of China's economic liberalisation since 1978. Due to their complex global networks of production and sales, TNCs create connections that tie local and national economies into the global economic system. TNCs have contributed to globalisation by:

- **Outsourcing** some parts of their businesses, usually administration and data processing, to third-party companies: Bangalore in India has become known as a location for TNC call centres and data processing.
- **Offshoring** some parts of their businesses to cheaper foreign locations, especially the special economic zones in Asian countries.
- Developing new markets: many TNCs that initially set up factories in Asia now sell their products there.
- Glocalisation: adapting brands and products to suit local market tastes. McDonald's is a classic example, which has adapted its products to suit the Indian market by not using beef, and offering spicy vegetarian products.

On the down side, TNCs have been accused of exploiting workers in the developing and emerging worlds by paying very low wages. Outsourcing jobs to the developing world can lead to job losses in developed countries. Local cultures and traditions can be eroded by TNC brands and western ideas.

## Switched off from globalisation

Most of the world is increasingly integrated into the globalised economy but some places remain 'switched off' and have only weak connections to other places. Table 4 summarises the reasons for this.

**Table 4** Switched-off places

| Political isolation | Physical isolation |
|---|---|
| North Korea has deliberately isolated itself from the rest of the world, shunning world trade and limiting the use of technologies such as mobile phones and the internet in pursuit of its own state ideology. | The Himalaya mountain countries of Nepal, Bhutan and Chinese Tibet are isolated by terrain and winter snow, limiting their connections to the outside world — although tourism is changing this. |
| **Economic isolation** | **Environmental barriers** |
| Rural parts of Sub-Saharan Africa, especially the Sahel, are dominated by a subsistence farming economy with food produced to eat not to sell. These places are also poor, and their capacity to create connections is limited. | Harsh desert climates, extreme polar cold and dense tropical forests all limit the development of transport and trade connections meaning continental interiors and polar regions are less well connected than coastal locations. |

# What are the impacts of globalisation for countries, groups of people and cultures, and the environment?

- Globalisation has benefited many people, but by no means everyone, thus creating winners and losers as the process has accelerated in both the developed and developing worlds.
- Migration is a key part of globalisation and it has affected some places much more than others, especially major cities.
- Cultural globalisation has created a 'western' global culture that may threaten existing and traditional cultures.

**Outsourcing** and **offshoring** are both ways of reducing business costs by moving parts of a TNC's business overseas. However, offshored parts are still owned by the TNC, whereas outsourced parts are not.

**Exam tip**

Learn some key facts and figures about TNCs to use in the exam as data always help give your answers added weight.

**Knowledge check 5**

Which was the world's largest TNC in 2016?

# The global shift to Asia

Many economists argue that the twenty-first century will be 'Asia's century'. Key evidence for this comes from the fact that the global economic centre of gravity (Figure 4) has shifted towards Asia in the last 30 years. This is a result of the **global shift** of industry towards Asia. In particular:

■ The shift of manufacturing jobs from Europe, Japan and North America to China and some other Asian countries such as the Philippines, Bangladesh and Vietnam.

■ The shift of service and administration jobs to India, especially the city of Bangalore.

> The **global shift** refers to the outsourcing and offshoring of industry, through foreign direct investment, to Asia since 1980.

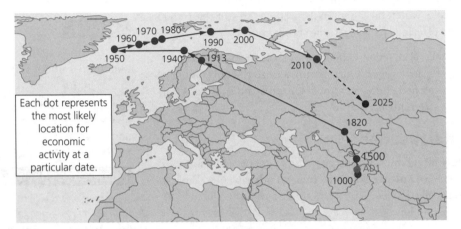

Each dot represents the most likely location for economic activity at a particular date.

**Figure 4** The global economic centre of gravity AD1 to 2025

Source: www.economist.com/blogs/graphicdetail/2012/06/daily-chart-19

# Environmental and social impacts of the global shift

On the face of it, this global shift would seem to be very beneficial to Asia but any rapid economic change tends to have costs and benefits, which are outlined in Table 5.

> **Knowledge check 6**
>
> In what year on Figure 4 was the global economic centre of gravity closest to North America?

> **Exam tip**
>
> Many exam questions will require you to assess the costs and benefits or advantages and disadvantages, rather than just describe or explain.

**Table 5** Costs and benefits of the global shift for Asia

| Benefits | Costs |
|---|---|
| Major investment in roads, ports, airports and power infrastructure; China built 11,000 km of new motorways in 2015 alone. | Urban sprawl and loss of productive farmland and forests as industry and cities expand to accommodate industry and worker housing. |
| A shift from informal, insecure employment to waged employment with a set income and some security. | New developments tend to be unplanned and sometimes poorly built, lacking key public services. |
| TNCs invest in training and skills development to improve workforce productivity, and some skills are transferable. | Pressure on natural resources, especially water supply as new factories and offices demand resources. |
| Major reductions in regional poverty due to employment; some 600 million Chinese were lifted out of poverty between 1992 and 2015. | Low wages, long working hours, lack of union representation and possible exploitation of workers. |
| As more people in formal employment pay taxes, local and national government invest in public services such as education and health. | Rapid loss of tradition such as local foods and dress as the pace of urban and industrial change is so rapid. |

It is clear that China has paid a heavy environmental price as a result of the global shift. Since 1980 it has undergone an industrial revolution similar to the one the UK underwent from 1770 to 1900. China's environmental issues include:

- Severe air pollution in cities like Beijing, where air pollution is regularly well above World Health Organisation safe limits.
- Beijing's 6 million cars and coal-burning power stations are the source of this pollution; close to 50% of all the world's coal is burned in China.
- Around 50% of China's rivers and lakes and 40% of its groundwater is polluted — so much so that it is unsafe to drink untreated.
- Over 20% of China is subject to desertification and severe soil erosion, which can create major dust storms.
- Combined with deforestation, desertification has forced many farmers off their land and into cities as the farmland has been over-exploited.
- The WWF reported in 2015 that almost half of China's land-based vertebrate species have been lost in the last 40 years as biodiversity has suffered as habitats have been destroyed.

Importantly, these environmental issues have human consequences as it is people who have to live within a polluted environment. Air pollution in northern China has been estimated to reduce life expectancy by nearly five years.

For developed countries the global shift has meant deindustrialisation. A positive aspect of this is lower industrial pollution levels: it could be said that the global shift has exported pollution to Asia. However, economic restructuring has caused a number of social and environmental problems in many former industrial cities in the developed world such as Detroit and Chicago, Essen and Dortmund, and Sheffield and Manchester:

- Declining populations: Detroit's population fell from 1.5 million in 1960 to 0.7 million in 2012 as a result of the closure of its car factories.
- Partly due to poverty Detroit has one of the highest violent crime rates in the USA with close to 2000 violent crimes per 100,000 people; in the UK post-industrial Middlesbrough has a high crime rate.
- About 4% of all land is Glasgow is derelict; this land is mostly disused factory sites and is often contaminated by industrial waste, making it costly to reuse.
- Unemployment is usually high in deindustrialised cities. In 2016 it was 6% in Pittsburgh USA and 9% in Hull — higher than the national average in both cases.

## Migration and globalisation

The connections created by globalisation have caused an increase in global migration. There is a greater 'churn' of people migrating for work than ever before. Probably the most significant form of migration is rural–urban migration. This feeds the growth of the world's megacities. In developing and emerging countries about 60% of urban growth is caused by rural–urban migration and 40% by high birth rates in cities (internal growth or natural increase). Figure 5 shows how some of the world's megacities have grown in recent decades. While some are growing slowly, developing world cities like Lagos and Karachi have very rapid growth.

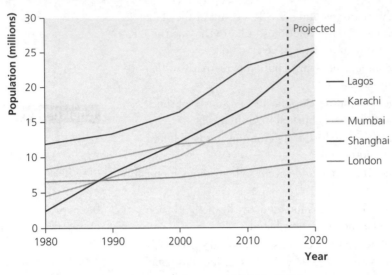

**Exam tip**

Learn some population growth data for contrasting megacities.

**Knowledge check 8**

Which city on Figure 5 has seen the most rapid growth since 1980?

**Figure 5** Megacity growth 1980–2015

In emerging and developing megacities rapid urban growth creates a number of social and environmental challenges, which are summarised in Table 6.

**Table 6** Social and environmental challenges of megacity growth

| Social challenges | Environmental challenges |
|---|---|
| Housing is in short supply, leading to the growth of slums and shanty towns that lack water, sewers and power supplies. | Sprawling slums at the city edge cause deforestation and loss of farmland and increase flood risk. |
| Poverty is rife, because wages are low and jobs are in short supply; many people have dangerous informal jobs. | Wood fires, old vehicles and industry mean air pollution levels are high. |
| Lack of taxes means city governments struggle to supply essential health and education services. | Rivers and lakes are polluted with sewage and industrial waste, making health problems worse. |
| Lack of water and sanitation means disease and illness are common in slums. | Critical resources, especially water, are in short supply because of soaring demand. |

Migration can also be international. Migrants are especially attracted to global hub cities. These migrants are of different types:

- HQs and offices of TNCs are often located in **global hubs**, so high-paid professional workers (lawyers, stock-market traders, bankers) are attracted to these places and this creates huge wealth.
- These global elite migrants often employ maids, drivers, nannies and gardeners.
- This attracts low skills migrants such as Indian and Bangladeshi migrants moving to the United Arab Emirates or Filipinos migrating to Saudi Arabia (in 2015, 27% of the UAE's population was from India).
- Further low skills, low wage migrants are used as construction workers for office and apartment blocks in global hubs.

Some cities, like London and New York, attract exceptionally wealthy migrants. An example is Russian oligarch billionaires (meaning wealthy business people) investing

A **global hub** is a city — like London, Dubai or New York — with an unusually high density of transport, business, political and cultural connections to the rest of the world.

in property in London and living there some of the time. This happens partly so the oligarchs can easily send their children to the UK's elite private schools, and partly to move money out of Russia and invest it in London property.

Migration has a number of costs and benefits which are summarised in Table 7.

**Table 7** Costs and benefits of migration

| | Economic | Social | Political | Environmental |
|---|---|---|---|---|
| **For the source country (origin of migrants)** | ✓ Remittances boost the incomes of families ✗ Loss of skilled and educated workers | ✓ Contact with a different culture ✗ Families are broken up as young males tend to migrate | ✗ Mass emigration can be viewed as a failure to provide for people at home | ✓ Reduces pressure on resources if the population is large |
| **For the host country (destination of migrants)** | ✓ Low wages workers fill skills gaps ✗ Some host population workers can't get jobs | ✓ Migrants can counteract an ageing population ✗ Demand for education, health and housing rises | ✗ Cultural tensions with the migrant population | ✗ Can lead to demand for more housing and therefore loss of greenspace; possible overcrowding |

# Emergence of a global culture

Globalisation has often been said to have spread a 'westernised' global culture which originates in North America and Europe. This is a culture based on:

- wealth creation, earning money in order to buy consumer goods and high levels of consumption
- private enterprise, where people own businesses rather than the government owning them
- success, which is measured by how wealthy you are and how much 'stuff' you buy
- fashion, technology and trends, which are important in western culture
- an attitude that the physical environment should be exploited for its natural resources to create wealth.

Western culture has spread by **cultural diffusion**. Migration dramatically increases cultural diffusion as migrants move and 'churn', so bring different people into contact with each other. Other factors are important too:

- Tourism brings people into contact with new cultures.
- TNCs spread their brands and products around the world.
- Global media organisations like Disney, CNN and the BBC spread a western view of world events.

Western culture can be viewed as having both positive and negative impacts on the physical environment and people. The spread of a western diet (high fat, high sugar, fast food based) is changing diets around the world, especially in Asian cities, with the spread of McDonald's, KFC and other fast food. This has been linked to rising obesity and diabetes in many emerging countries. A fast-food, consumer culture

**Remittances** is the money migrants send back home to help their families in the source country.

**Knowledge check 9**

What are remittances?

**Cultural diffusion** is the exchange of ideas between different people as they mix and interact as a result of globalisation.

is also very wasteful in terms of resources such as discarded fast food packaging and fashion items worn only once or twice. This can be linked to deforestation and excessive water use in industry, as well as air and water pollution. On the other hand western culture has tended to improve opportunities for some traditionally disadvantaged and discriminated against groups such as women, the disabled and LGBT groups. Global media coverage of the Paralympics, Gay Pride marches and high profile cases of sex discrimination may help erode discrimination and prejudice in developing and emerging countries.

## Opposition to globalisation

The spread of western culture is strongly opposed by some groups, broadly called the anti-globalisation movement. Protest groups such as Occupy Wall Street and the Global Justice Movement argue that globalisation has:

- dramatically increased resource consumption through exploiting the natural environment, leading to problems like deforestation, water pollution, global warming and biodiversity loss
- exploited workers, especially in emerging countries, who suffer low wages, dangerous working conditions and lack any form of union representation
- passed political and economic power into the hands of TNCs and uncaring governments, at the expense of ordinary people
- created increased inequality, i.e. a small group of very rich, powerful people (sometimes called the '1%'), at the expense of others
- caused cultural erosion, meaning that traditional lifestyles are degraded by the spread of western culture, and local dress, art and architectural styles are lost.

Opponents of globalisation also point to its impact on traditional cultures and lifestyles. Because nowhere is untouched by globalisation the number of people able to live isolated, traditional lifestyles is now very small. Arctic Inuit, tribal groups in Papua New Guinea and Amazonia, and mountain people in Nepal and Bhutan now all experience tourism and exposure to global media. Their traditional foods, music, language, clothes and social relations are all being eroded, or else being turned into a 'show' for tourists.

# What are the consequences of globalisation for development and the environment?

- Globalisation has led to increased development in some countries, but has also widened the gap between rich and poor in some cases.
- Not all people, or all countries, view globalisation as a positive development and some places have attempted to limit its impact.
- There are alternatives to globalisation, which focus on local communities and local production, and attempt to be more sustainable socially, economically and environmentally.

## Globalisation and the development gap

The development gap refers to the difference between the richest and poorest people. This can be:

**Knowledge check 10**

Name one health impact of changing diets in Asia.

**Exam tip**

The word 'culture' can be challenging to interpret in the exam. It means the commonly held views, traditions and lifestyles of a group of people. It is different from the word 'social', which is more about the health and wellbeing of people.

- between countries, e.g. in 2015 people in Luxembourg had incomes of $105,000 per year compared to only $220 in South Sudan
- within countries, e.g. in China's coastal cities per capita incomes are over $10,000 whereas in the rural west they are under $2000.

Measuring the gap between rich and poor is not easy. Geographers use single measures like life expectancy or GDP per capita because they give an easy to use and understand 'headline' measure of **development**. However, single measures are not very accurate.

Instead, composite indices are use. They combine several data points into an index. The most well known is the combination of life expectancy, income and years in education used to produce the Human Development Index (HDI). The Gender Inequality Index (GII) combines the reproductive health of women, their participation in the workforce and empowerment (women in higher education and politics) to measure gender aspects of development. These indices focus on social development as well as economic development and are usually viewed as a better reflection of development progress. Table 8 compares development data for three countries.

**Development** means progress, usually in social (health, diet, life expectancy, housing quality) and economic (income) terms.

**Exam tip**

Learn some development indicator data for a few countries to use as evidence in your answers.

**Table 8** Comparing development indicators

| Country | Single measures | | | Composite indices | |
|---|---|---|---|---|---|
| | Nominal GDP per capita (US $) | GDP per capita PPP ($) | % of people employed in farming economic sector | HDI (1 = best, 0 = worst) | GDI (0=best, 1 = worst) |
| Sweden | 48,900 | 47,900 | 2 | 0.85 | 0.05 |
| Mexico | 11,200 | 17,500 | 13 | 0.58 | 0.37 |
| Haiti | 800 | 1,750 | 38 | 0.29 | 0.80 |

Purchasing Power Parity (PPP) GDP per capita has become a popular way of comparing economic development between countries because unlike nominal GDP it takes into account the cost of living within countries. We can also measure development using environmental indicators such as WHO air pollution levels. However, these tend to be local, i.e. for specific cities, so can't be used to compare countries.

**Knowledge check 11**

Name a composite index that measures development.

## Widening income inequality

A 2016 report from Oxfam stated that the wealth of the world's richest 1% of people is equivalent to the wealth of the other 99%. This degree of income inequality is not new, but it may have become starker in the last few decades. Within countries income inequality is measured using the Gini Coefficient (Figure 6) with income divided into quintiles (20% intervals) plotted as a Lorenz curve. Figure 6 shows that Haiti is the most unequal country as the richest 20% of people have 65% of the wealth, compared to under 40% in Sweden.

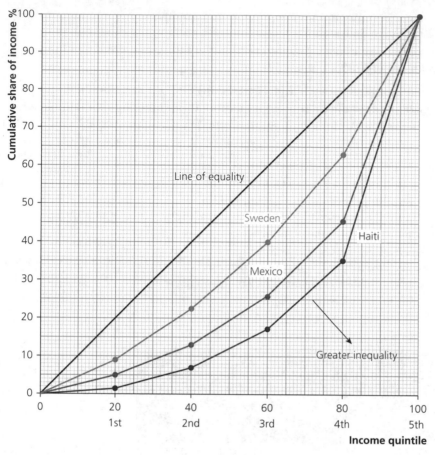

**Figure 6** Lorenz curve Gini Coefficient plots for three countries

It is important to recognise that Sweden, Mexico and Haiti have all experienced globalisation to some degree, but the outcome is that Mexico and Haiti are more unequal than Sweden. It is hard to generalise but globalisation has created winners and losers as shown in Table 9.

**Table 9** Globalisation's winners and losers

| Winners | Losers |
|---|---|
| There were about 1800 billionaires worldwide in 2016; most have made their wealth through ownership of global TNCs. | Isolated, rural populations in Asia and Sub-Saharan Africa where subsistence farming still dominates and global connections are thin. |
| Developed countries have proved very good at maintaining their wealth, despite the rise of countries like China. | Workers (especially male ones) in old industrial cities in the developed world who have generally lost jobs. |
| The rising middle class of factory and call centre workers in Asia, whose incomes have risen as they have gained outsourced and offshored jobs. | Workers in sweatshop factories in emerging countries; they suffer exploitation (but may still be better off than in the rural areas they migrated from). |
| People who work for TNCs in developed countries who have a high income and reasonable job security, although lead high-stress lives. | Slum dwellers in developing world megacities like Lagos, as the reality of urban life is often much worse than they expected. |

**Knowledge check 12**

What does the Gini Coefficient measure?

**Knowledge check 13**

How many billionaires were there worldwide in 2016?

**Exam tip**

Good answers will recognise that some people, such as factory workers in China, are both 'winners' and 'losers' from globalisation, i.e. acceptable incomes but poor working conditions.

The environmental impact of development and globalisation is often measured using **ecological footprints**, whereas one way of measuring economic development is using income per capita. Figure 7 compares trends in both for China and Sweden since 1970. Figure 7 shows that:

- Sweden's income per person has grown hugely, but its ecological footprint has not.
- This suggests that economic development in Sweden has not affected the quality of the environment and that environmental management maintains biodiversity, water and air quality.
- China's ecological footprint has steadily risen (but is lower than Sweden's).
- Since 2001, rising Chinese incomes correlate with very large increases in ecological footprint.
- This suggests that economic development in China has very high environmental costs.

In summary: some countries can take advantage of globalisation without damaging their environment while others cannot.

An **ecological footprint** is a measure of the resources used by a country or person over the course of a year, measured in global hectares.

**Figure 7** Trends in income per capita and ecological footprints

## Tensions caused by globalisation

For a large number of countries a significant part — or even majority — of their total population consists of immigrants. In 2015, 84% of the UAE's population was immigrant, 29% in Switzerland, 14% in Germany and the USA and 11% in the UK.

Globalisation has contributed to large immigrant populations and there are now large **diasporas** from many countries resident in other countries. Several factors have increased the pace of migration:

- Open borders to migration within the EU since 1995.
- FDI, encouraging TNC workers to move overseas.

A **diaspora** is the name given to the dispersal of a population overseas; since 1700 about 10 million Irish have emigrated overseas creating the Irish diaspora.

- Deregulation of some jobs markets, allowing foreign qualified workers.
- Humanitarian crises, like the Syrian civil war and war with Islamic State, which have seen large numbers of refugees flee to Europe since 2011.

Most EU countries, as well as many other developed countries, now have culturally mixed populations. Large-scale migration is not without costs. Migrants need housing, jobs, education for their children and other services. At a certain rate of immigration all of these will come under strain and this risks a rise in tensions with some of the host country population who may view the migration as 'too many, too fast'. There is evidence in Europe that migration has increased social and political tensions and even led to a rise in extremism:

- The UK 'Brexit' vote in 2016 to leave the EU had the scale and pace of immigration as a key area of debate.
- Anti-immigration political parties have been rising in popularity since 2010, for example, UKIP in the UK, the Front National in France, the Dutch Party for Freedom, and Freedom Party of Austria.
- In 2014, 51% of Swiss voted in favour of stopping mass immigration in a national referendum.
- Even in the USA, a country of immigrants, the benefits of migration from Mexico and elsewhere have been questioned.

Few people in developed countries vote for extremist political parties that seek to ban immigration or return immigrants to their home country. However, such sentiments are more common than they once were.

## Controlling the spread of globalisation?

Some countries have attempted to limit the impact of globalisation using government policy:

- The internet is banned in North Korea, because the Supreme Leader Kim Jong-un does not want his people to have access to 'western' ideas.
- In China, the internet was very widely used by 52% of the population in 2016, but it is censored; some searches for politically sensitive topics get no results because the Chinese Communist Party seeks to prevent 'unhelpful' discussion.
- Since 2010 the UK has sought to reduce immigration using a points system, but with only limited results because EU immigration cannot be controlled.
- Other countries like Australia also use **points-based immigration systems** to match immigrants to actual economic needs and job vacancies.
- Trade protectionism is still common: oil exports are banned in the USA so all domestically produced oil must be used in the USA; India restricts foreign companies investing in its retail sector to protect Indian small shopkeepers from competition.

These restrictions are generally the exception rather than the norm, as most countries operate within a system of global free trade and open access to information, and broadly welcome the economic advantages of globalisation. Table 10 summarises how one indigenous group, the First Nations of Canada, attempts to retain its cultural identity and prevent it being eroded by cultural globalisation.

**Knowledge check 14**

What is meant by the term 'diaspora'?

**Exam tip**

Don't over-emphasise extremism in answers about immigration. In almost all countries it is a minority view.

**Points-based immigration systems** award points to potential immigrants based on education, skills, language proficiency and other criteria so that migrants are matched to a country's needs.

**Table 10** Preserving the cultural identity of the First Nations of Canada

| The First Nations are the original population of Canada, existing before European immigration. They consist of Indian bands such as the Cree and Lenape. | | |
|---|---|---|
| An Assembly of First Nations promotes the rights and needs of First Nations at national level within Canada. | Within Indian Reservation territories, bands are largely self-governing, allowing them to make key decisions about their future. | There are about 100 First Nation and Inuit Cultural Education Centres funded by the Canadian Government to help preserve and develop First Nation cultures and traditions. |
| After decades of being taught to be 'Canadian' in boarding schools, modern First Nation schools teach native languages and traditions. | Festivals and other meetings help preserve the First Nations tradition of oral histories and other traditions. | To some extent, tourism helps preserve some aspects of First Nation culture but also risks diluting it. |

## Ethical and environmental concerns

Globalisation has led to a number of ethical and environmental concerns which are quite widely held:

- fears that consumer goods have been made using exploited labour
- concerns that imported food products like tea, coffee, bananas and cocoa do not provide their farmers with a decent income due to low prices
- concerns that consumer goods use excessive resources during their production, packaging, transport and use
- worries that our consumer culture is contributing to global warming as ecological footprints rise.

These concerns can be illustrated by something as simple as a pair of jeans bought in the UK but made in Bangladesh (Figure 8).

To grow the cotton to make the denim fabric uses 13,000 litres of water

Fertilisers and pesticides from cotton farms leak into rivers and groundwater

Dyes and chemical fabric treatment waste from factories poison rivers

Cotton farmers in Africa live on less than £1 per day

Bangladesh's 3.5 million textile workers earn about £25 per month

Many textile workers work 14-hour days in appalling conditions

Jeans need to be transported by ship from Bangladesh, which burns fuel-oil and releases greenhouse gases

Most cheap jeans last 12–18 months before being discarded for the latest fashion

**Figure 8** The hidden costs of a pair of jeans

**Ethical issues** are those which have a moral dimension, and concern whether something is acceptable. Child labour is an example of something most people find unethical/morally unacceptable.

**Localism** is the idea that food and goods should be grown and made locally, supporting local jobs and reducing transport, rather than being sourced globally.

There are several different responses to the social and environmental **ethical issues** raised by globalisation and globalised consumer products. A key response in developed countries has been a move towards **localism**, i.e. buying local products, trying to

# Content Guidance

trade with other local businesses and building local community movements around sustainability issues. These are summarised in Table 11.

**Table 11** Localism

| Transition towns | Fair Trade |
|---|---|
| Founded in 2006 the non-governmental organisation (NGO) 'Transition Network' encourages towns to grow their own food in community gardens (not import it) and reduce energy used in transport, e.g. cycling and recycle waste / reuse materials. Some towns like Totnes, Exeter and Stroud even have their own local currencies to encourage local trade.<br><br>These initiatives are small scale, but some elements like 'grow your own' could have a big impact if widely adopted and promoting local sourcing became more widespread. | Fair trade — rather than free trade — pays farmers of cocoa, cotton, tea and coffee in developing countries a guaranteed price for their produce plus a 'fair trade premium' payment. This attempts to reduce the inequalities of global trade.<br><br>The aim is to make incomes sustainable for farming families, and use some of the additional money to support community facilities like wells, schools and clinics.<br><br>The downsides of fair trade are that the extra income is small, and fair trade products are more expensive for consumers. |
| **Ethical consumption schemes** | **Recycling** |
| Founded in 1993 in Germany the NGO FSC (Forest Stewardship Council) uses its FSC logo on wood products that are sourced from sustainable forests thus helping consumers ensure that products are not contributing to environmental degradation.<br><br>Its criteria include that forestry must respect the land rights of indigenous people and that forestry workers are well treated and paid.<br><br>FSC has become well known globally but has been criticised for being too brand-focused. | Local councils in the UK play a key role in reducing waste and ecological footprints through recycling and councils' waste collection services.<br><br>Recycling of household waste increased from 17% to 44% between 2003 and 2013 but this was still some way behind the 65% achieved in Germany.<br><br>Recycling does reduce waste, but different councils have different schemes with different results and reducing packaging might be a better way forward. |

**Knowledge check 15**

What does the FSC logo on products indicate?

**Exam tip**

You need a range of examples of ethical and environmental initiatives that aim to reduce some of the negative consequences of globalisation.

## Summary

- Globalisation has its origins several centuries ago, but the process of increased global connections and flows has accelerated in recent decades.
- Transport and communications technology, and international organisations, governments and TNCs all have important roles to play in globalisation.
- Some places are more globalised than others, with a number of places being physically or politically 'switched off' from the rest of the world.
- Asia is at the centre of twenty-first-century globalisation but this has brought with it environmental costs, whereas in North America and Europe deindustrialisation has occurred.
- Globalisation has fuelled migration, both from the countryside to cities and internationally between countries, which has both benefits and costs.
- There is increasing evidence of a global culture as western values spread worldwide but this has both costs and benefits.
- Levels of economic and social development vary markedly around the world, although development progress has been made in almost all places.
- Globalisation, especially migration and cultural globalisation, has not been welcomed everywhere and some governments and groups have tried to restrict its influence.
- There are alternatives to globalisation including localism and ethical consumption, but these have not been adopted in a widespread way.

# ∎ Shaping places

## Regenerating places

### How and why do places vary?

- Economic activity and job type varies from place to place.
- Over time, the economic and population characteristics of places change for a number of reasons which are national or local.
- Places are also influenced by global change which affects the lives of groups of people who live there, such as students and others.

This section of the guide compares two contrasting places in the UK: Reading and Middlesbrough. You will have studied your own two places. You should relate the places you have studied to the themes in the content guidance below. You could also use Reading and/or Middlesbrough as additional case studies to your own.

### Classifying economies

Despite the UK being a relatively small country there is considerable variation in economic activity between places. **Economic sectors** vary from place to place reflecting different economic processes that have affected places over time. In general:

- Rural areas have more primary employment in farming, mining, quarrying and fishing; this tends to be low paid, manual work.
- There is more secondary employment in northern cities such as Manchester, Sheffield and Glasgow, but this has declined over time.
- Tertiary, or service sector, jobs are concentrated in urban areas but these vary from cleaners on minimum wage to very high paid professionals like lawyers.
- Quaternary jobs in research and development and hi-tech industries are found in London and the South East.

Table 12 shows job type in Reading (southern England) and Middlesbrough (northeast England). These are similarly sized urban areas. Notice that Reading has more than twice the number of professionals, but far fewer skilled trades, caring and leisure workers and people employed in manual work.

**Table 12** Employment in Reading and Middlesbrough in 2015

| Pay and education | Percentage of people who are: | Reading | Middlesbrough |
|---|---|---|---|
| Highly paid, high level of education | Management | 9.2 | 7.5 |
| | Professional (lawyers, doctors) | 28.7 | 13.6 |
| | Professional & technical | 16.1 | 12.1 |
| | Administrative | 10.2 | 10.4 |
| | Skilled trades | 8.8 | 11.7 |
| | Caring, leisure & other services | 7.6 | 12.2 |
| | Sales & customer services | 5.3 | 8.1 |
| | Process plant & machine operators | 4.5 | 6.7 |
| Low pay, low level of education | Manual work | 9.6 | 16.7 |

The four **economic sectors** are primary (farming, mining), secondary (manufacturing), tertiary (retail services, office work) and quaternary (scientific research, ICT).

**Knowledge check 16**

Which economic sector does someone working in a factory making mobile phones work in?

**Exam tip**

You need to know some data and statistics on economic activity in your two chosen places.

Table 12 suggests Reading is a more economically successful place than Middlesbrough. Further data confirms this. In 2015 average hourly pay for a male worker in Middlesbrough was £12.50 and in full-time employment a male worker could expect to earn £532 a week. In Reading the figures are £14.80 and £605.

Figure 9 breaks down people in the two places by economic activity. It shows that 66% of people in Reading are economically active (working) compared to 54% in Middlesbrough, where part-time work (which pays less) is much more common. **Gross Value Added** (GVA) for the two places is also very different being £34,000 per person in Reading and only £17,000 in Middlesbrough.

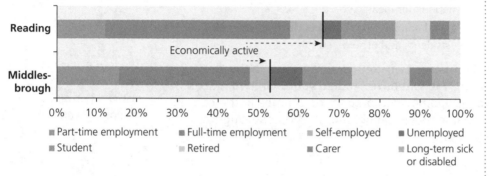

**Part-time employment**  **Full-time employment**  **Self-employed**  **Unemployed**
**Student**  **Retired**  **Carer**  **Long-term sick or disabled**

**Figure 9** Economic activity in Middlesbrough and Reading in 2015

The different economic characteristics of Reading and Middlesbrough have a number of causes including differences in education and pay:

- In 2014, 22.5% of people in Middlesbrough had no educational qualifications, versus 11.5% in Reading; 19% had a university level qualification in Middlesbrough but the figure was 43% in Reading.
- Middlesbrough's manual workers earn about £350 per week, whereas Reading's professionals earn over £700.

These differences have led to a number of consequences:

- Temporary, low paid and 'zero-hours contract' work is more common in Middlesbrough, meaning people have lower job and income security.
- Perhaps surprisingly there is a difference in life expectancy; if you were a male born in Middlesbrough in 2014 you would be expected to live to age 77, but if you were born in Reading that figure is 81.
- In 2016 the annual uSwitch **Quality of Life** Index ranked Berkshire (where Reading is located) 6th out of 138 UK regions, but South Teesside (Middlesbrough) was ranked 129th.
- Notice on Figure 9 that health — measured by the percentage of long-term sick and disabled — is very high in Middlesbrough compared to Reading.

**Exam tip**

If one of your places is a less successful place, be careful not to overstate its problems.

**Gross Value Added** (GVA) is a measure of economic output; it is the value of goods and services produced per person.

**Quality of life** is usually measured using a composite index. The uSwitch Index combines housing affordability, energy costs, broadband availability, average incomes, crime rates and other measures to rank UK regions.

**Knowledge check 17**

What is the difference in male life expectancy between Reading and Middlesbrough?

# Changing function and characteristics

Explanations for differences between places focus on economic and population change over time. Very few places are 'static' so change affects places continually. The places we live in have a range of economic functions which reveal themselves in the land use of urban and rural places:

- Administrative: council offices, schools and other public services like clinics and hospitals.
- Commercial: offices of service industries such as legal services, accountants.
- Retail: shops that range in size from small to malls.
- Industrial: factories, warehouse and distribution centres.

Towns and cities have a mix of these functions, but this changes over time based on economic health:

- Industrial land use in Middlesbrough has declined since the 1970s due to factory closures.
- In Reading commercial functions have grown due to the success of the area's service sector, plus the location of some companies in the quaternary industrial sector such as Oracle, Microsoft and Intel.

By now, you should have an image of Reading as a 'successful' place and Middlesbrough as less successful. This degree of success is reflected in the demographic characteristics of the two places shown in Figure 10. Reading has a much higher proportion of people in the 20–44 age category suggesting young professional workers, whereas there are more older and retired people in Middlesbrough. The **ethnic composition** of the two places is also different. In 2011, 65% of Reading's residents were white British compared to 86% in Middlesbrough. Greater ethnic diversity in Reading suggests a more successful place that has attracted economic migrants.

**Ethnic composition** is the ethnic group make-up of a popualtion. In the UK the main groups are white, Asian and black. Ethnicity is different from nationality (country of origin).

**Exam tip**

Population age groups, and growth or decline, are very important to understanding places so make sure you learn some data on each.

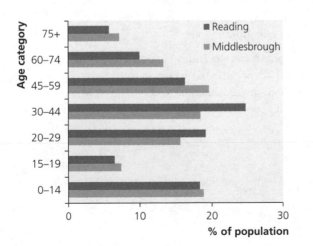

**Figure 10** The age profile of Reading and Middlesbrough in 2011

There are many reasons that can be used to explain the changes and relative success of Middlesbrough and Reading and these are summarised in Table 13.

**Table 13** Explaining the success of places

|  | Middlesbrough | Reading |
|---|---|---|
| **Accessibility** | Not on the UK motorway network or a mainline train route.<br>A long distance north of London, the economic core. | Located on the M4, just west of London and with very good rail links.<br>Benefits from its proximity to London. |
| **Connectedness** | The local airport, Durham Tees Valley, is closing down.<br>Limited higher education opportunities. | Close to the global hub airport of Heathrow.<br>Very close to many major universities, including Reading. |
| **History** | A centre for industrial revolution mining, shipping, engineering, steel and petrochemicals; however, many of these industries have now closed. | Part of the M4 growth corridor west of London, it has become a centre for **footloose industry** and services since the 1970s. |
| **Planning** | Almost a forgotten corner of the North East, it has never benefited from government regional investment. | A London 'overspill' town beyond London's **greenbelt**; close to rural areas, it is an attractive place to live. |

## Measuring change

Data can provide a picture of these two places. Change can be measured using employment trends, demographic change and changes to deprivation levels. Most tellingly, Middlesbrough's population of 146,000 in 1991 had fallen to 138,400 by 2011. In contrast, Reading's population grew from 136,000 in 1991 to 155,000 in 2011. Population growth or decline is a key indicator of how economically successful places are.

Reading, and nearby Bracknell, had gained over 40,000 **digital economy jobs** by 2016, whereas in 2015 Middlesbrough was in the news because of the closure of the Teesside Steelworks with the loss of 3200 direct and indirect jobs.

A useful way to understand places in detail is to examine the Index of Multiple Deprivation (IMD). This is an attempt to quantify deprivation in England. It uses seven data domains (Figure 11) which are weighted towards income and employment. The IMD is a very 'fine grained' index because it splits England into nearly 33,000 small areas with about 1500 people each.

| Income | Employment | Education | Health | Crime | Barriers to housing and services | Living environment |

**Figure 11** The seven domains of the Index of Multiple Deprivation

**Footloose industries** are those that can be located anywhere; they are not tied to locations by natural resources or fixed infrastructure.

**Greenbelts** are land surrounding cities that cannot be built on, usually farmland. Development sometimes 'leapfrogs' the greenbelt, benefiting places just beyond the greenbelt edge.

### Knowledge check 18

What happened to Middlesbrough's population between 1991 and 2011?

**Digital economy jobs** are those in industries like mobile technology, ICT, software design and app development.

Using the 2015 IMD data for 326 local council areas in England, Middlesbrough was the 7th most deprived area in England whereas Reading was 147th with less than half the level of **deprivation** in Middlesbrough. Reading's rank may seem a little high, i.e. deprived, but the very lowest levels of deprivation tend to be found in rural areas and commuter belt towns and villages in the South and South East. All large towns and cities have some degree of deprivation and Reading is no exception.

## Past and present connections

Places, almost wherever they are, are influenced economically and socially by regional, national, international and global forces which affect them in both positive and negative ways. These forces are summarised for Reading and Middlesbrough in Table 14.

**Table 14** Forces shaping Middlesbrough and Reading

|  | Reading | Middlesbrough |
|---|---|---|
| **Global** | Global brands like Verizon, Oracle, Microsoft, Cisco, PepsiCo and Vodafone have all located in Reading International Business Park. Many globally known tourist sites (Windsor Castle, Ascot racecourse) are close by. | The global shift of manufacturing industry has led to factory closures and a loss of jobs. Global competition has made its steel and petrochemicals industries less profitable. |
| **International** | Close to the economic core of the EU, and within the EU single market. The EU is easily accessible by air, road (Channel Tunnel) and ferries. | Middlesbrough has received EU economic development funding as a 'Transition Region', but less than areas like Cornwall and Wales. |
| **National** | High transport spending in the South and South East has provided connections: the M4, Heathrow airport, the M40 and M3. | It is located in a 'cut-off' corner of the North East, too far east of the A1 trunk road and east coast mainline railway. |
| **Regional** | London's greenbelt has made Reading a nearby alternative where development is allowed. Migrants are attracted to Reading, being close to London and close to their UK entry point. Within the M4 corridor ('Silicon Valley'), it is the preferred location for hi-tech industry in the UK. Many people live in the area, but commute to London. | Iron ore deposits, which were in part the origin of Middlesbrough's industrial growth, were exhausted decades ago. There are poor road connections to nearby cities such as Leeds and Newcastle, and very poor rail connections. It is close to the North York Moors National Park, but not close enough to benefit from tourism. |

Fairly or unfairly, all places have an image that they project and this shapes people's perceptions of the place as either positive or negative. This image can also have an effect on people in the place. Their **identity** may be affected if they perceive they are living in an area that has a positive or negative image. Figure 12 shows images of both places. They are not that different, but the Reading image could be perceived as more modern, busy and as a place where there is lots to do. The Middlesbrough image is more industrial and less attractive.

**Deprivation** means lacking things that are considered normal by society, such as a job, decent income, warm secure housing or access to healthcare. Multiple deprivation means lacking several of these.

**Knowledge check 19**

How many data domains are combined into the Index of Multiple Deprivation?
......................................

**Exam tip**

Scale (global, national, etc.) is an important concept. Make sure you know how forces at different scales have shaped your chosen places.
......................................

**Identity** refers to people's feeling and perceptions, and their shared beliefs, traditions and ways of life. It can create a sense of community and feeling part of a wider group of similar people.

**Figure 12** Images of Reading (top) and Middlesbrough (bottom)

Students, other young workers and migrants are affected by these images and perceptions:

- Young people may feel they want to leave a place with a poor image.
- People are attracted to places with positive images.
- There are likely to be more job opportunities in places with attractive images, because companies, like people, are attracted to them.

Since 2010, the UK Government has attempted to measure 'national wellbeing' by conducting a survey asking people about how they feel about their lives. Results from 2015 for Reading and Middlesbrough are shown in Table 15. The results are not dramatically different, but in all cases people in Middlesbrough answered low/medium more than people in Reading. This might give an insight into how people perceive their place.

**Table 15** National wellbeing survey results 2015

| How do you feel about: | Low/Medium | | High/Very high | |
|---|---|---|---|---|
| | Middlesbrough | Reading | Middlesbrough | Reading |
| Life satisfaction | 24% | 17% | 76% | 83% |
| Life is worthwhile | 20% | 15% | 80% | 85% |
| Happiness | 29% | 23% | 71% | 77% |

# Why might regeneration be needed?

- There are major economic and social differences between successful regions and those that need regeneration.
- There are variations in how attached people are to the places they live in, and views about regeneration can lead to conflict.
- The need for regeneration can be evaluated using a range of different data.

## Economic and social inequalities

There are many economically successful urban and rural regions. These places attract people and investment, but they are not free of problems. A good example is Santa Clara County in the San Francisco Bay area of California, USA. This is the original 'Silicon Valley' and hosts the HQs of Apple, Hewlett-Packard, Adobe and eBay in the cities of Cupertino, Palo Alto and San Jose. There are many indicators of this region's success:

- Population in 1990 was 1.5 million, growing to 1.9 million by 2015.
- Roughly the same size as the UK county of Kent, it has an annual GDP of $180 billion (similar to the Czech Republic).
- An average detached house in Santa Clara costs over US$ 1 million.
- Average household income in 2014 was $89,000.

Figure 13 shows that Santa Clara County has a very ethnically diverse population. It has attracted highly educated migrants from across the USA and internationally. In 2014, 198,000 immigrants gained residency or permission to work long-term in California, more than any other US state and about 20% of the total for the USA.

**Knowledge check 21**

Which are the largest ethnic groups in Detroit and Santa Clara County?

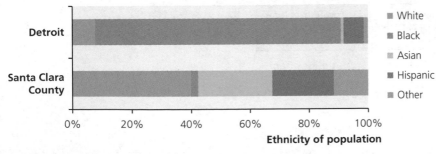

**Figure 13** Ethnicity in Santa Clara County and Detroit 2014

In contrast Detroit, a city in the state of Michigan, USA is declining. Once the centre of American car manufacturing, 'Motor City's' population fell from 1.5 million in 1970 to only 680,000 by 2015 — a huge drop of over 50%. Detroit is in the USA's **rustbelt** and has suffered from the effects of deindustrialisation.

- Average household income in Detroit was about $25,000 in 2015, half the national average and more than $60,000 lower than in Santa Clara County.
- By 2014, two-thirds of Detroit's' residents could not afford basic needs like food and fuel and the poverty rate was 38%.
- Life expectancy in parts of Detroit is just 69 years, and less than 30% of students graduate from high school.
- In 2014 Detroit had the second highest murder rate of any US city.
- Average house prices in Detroit are about $40,000 and it is estimated that in 2015 there were 30,000 abandoned homes and 70,000 other abandoned buildings.

**Exam tip**

You need to be able to use data to show why some places need to be regenerated.

In terms of **ethnicity** (Figure 13) Detroit is dominated by Black African Americans. This is because they are the lowest income group left behind when other groups — generally more skilled and better educated — have migrated from Detroit as it has declined.

Many industrial cities like Detroit and Cleveland in the USA, or Newcastle and Hull in the UK, entered a **spiral of decline** as a result of deindustrialisation, or negative multiplier effect. This is shown in Figure 14. Regeneration aims to stop, and then reverse, this decline.

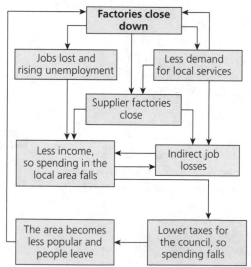

**Figure 14** The negative multiplier effect

The **rustbelt** is an area of industrial decline in the northeast USA including Buffalo, Cleveland, Detroit and Chicago.

**Ethnicity** refers to groups of people who share a common culture, ancestry, language and traditions — and often religion. Race (racial group) is based on physical/genetic characteristics.

A **spiral of decline** is a hard-to-stop loss of jobs, people and local tax revenues that leads to further losses and greater decline.

Santa Clara County and Detroit are in some ways extremes, yet they share some interesting problems:

- Detroit has severe shortages of public sector workers (teachers, nurses) because most have simply moved away to better places.
- Santa Clara has shortages of skilled workers, because living costs are so high, i.e. house prices and commuting costs.
- Increasingly Santa Clara is short of space to build new offices, malls and homes because demand is so high.
- Demand is so low in Detroit that huge areas of the city are simply abandoned.

Rural areas can suffer similar issues to urban ones. A number of changes have affected all rural areas in the UK:

- Falling employment in farming and other primary sector employment like mining, quarrying, fishing and forestry.
- Often this is a result of mechanisation, i.e. bigger, more efficient and sophisticated machines doing the work once done by human labour.
- Migration of young people out of rural areas for education and then employment.
- **Ageing rural populations.**
- Decline in rural services especially post offices, banks, petrol stations and some public services.
- Even in rural areas with population growth, services can decline because many people increasingly drive to use services in nearby towns.
- A shift in economic activity towards services, especially leisure and tourism, but this only benefits popular areas accessible from major towns and cities.

Very few rural areas in the UK have seen their populations decline but an exception is West Somerset. Its population fell by about 500 to 34,400 between 2001 and 2011. In contrast the population of Mid Devon, next door to West Somerset, increased from 69,900 in 2001 to 77,800 between 2001 and 2011. Table 16 contrasts these two rural areas.

**Ageing populations** are ones with an increasing percentage of over-65s, who are retired and dependent on pensions; they place high demands on public services especially healthcare.

**Table 16** West Somerset and Mid Devon

| West Somerset | Mid Devon |
|---|---|
| Has the oldest average age of any local council area in England, at 52, and has suffered from young people leaving | Has benefited from the growth of holiday homes and in-migration of retired people; there are some commuter villages |
| Has no motorways, higher education provision or rail connections, making economic development difficult | Very accessible via the M5 motorway and the Great Western railway. |
| Despite being close to Exmoor, West Somerset has a limited amount of tourism development | Has several tourism hotspots, including the Grand Western Canal and Exmoor National Park |
| In the 2015 IMD it ranked as the 56th most deprived local authority, high for a rural area | Ranked 156th most deprived local authority area in the 2015 IMD |

**Knowledge check 22**

Which group of people have tended to migrate out of UK rural areas in recent years?

**Exam tip**

Many exam questions will require you to compare areas, so you need several examples not just one big case study.

## Variations in inequality

In both rural and urban areas there are significant **inequalities** between places that are often spatially very close. This is especially true in urban areas where very deprived areas can often be found next to areas with almost no deprivation. Figure 15 shows deprivation levels in Reading. These are national data levels, so an area on Figure 15 in the 10% most deprived category is within the 10% most deprived areas in England, not just in Reading. Notice that:

■ Areas close to the town centre Central Business District (CBD) to the south and west are most deprived.

■ Northern areas of Reading are the least deprived.

■ Some areas of high deprivation are towards the edge of the urban area.

■ In some places very highly deprived areas are right next to least deprived areas.

> **Inequality** means a difference between two or more places, or groups of people. It could be measured in terms of average income or access to services. The term is often used to imply a difference that is 'unfair'.

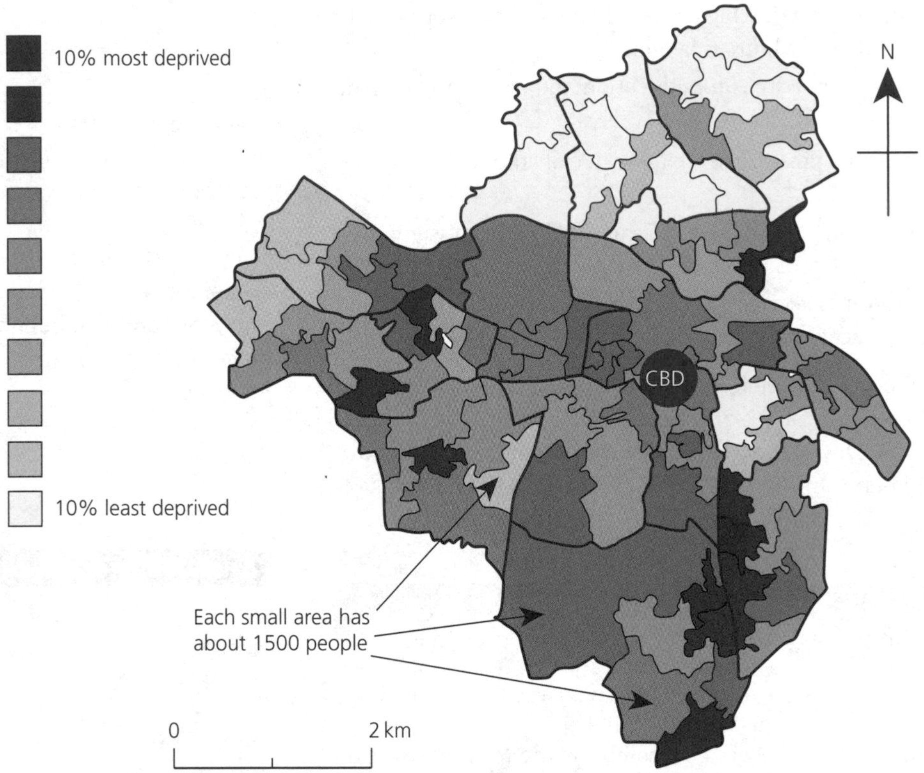

**Figure 15** IMD 2015 deprivation levels in Reading

In Reading, as in other places, areas that may need regeneration can be found right next to areas that have no need of it at all:

■ Rich **gated communities** can be found right next to '**sink estates**' in cities and towns.

■ In rural areas, successful, prosperous commuter villages might be only a few miles from less attractive rural villages suffering population decline and service deprivation.

> **Gated communities** are wealthy residential areas that are fenced off and have security gates and entry systems. They are increasingly common in the UK.

# Lived experience

People who live in areas in need of regeneration know they live in deprived, run-down areas but may not be in a position to do much about it, or want to. This is the idea of engagement — the extent to which communities are prepared to get involved in pushing for change. The most common way to measure engagement is to consider **election turnout**.

In the 2015 UK general election 66.1% of voters who could vote, did. There were large variations in turnout:

- 78% of people over 65 voted, compared to only 43% of people aged 18–24.
- Locally, turnout ranged from 51% in Stoke on Trent Central to 82% in East Dunbartonshire in Scotland.
- Turnout is lowest in deprived inner cities, and highest in wealthy suburbs and commuter belt rural areas.
- The proportion of men and women who chose to vote was very similar.
- Professional and manager turnout was 75%, whereas among manual workers it was only 57%.
- Only 55% of ethnic minority groups chose to vote.

In local elections, to elect local councillors who run councils on a day-to-day basis, turnout is much lower, only 36% in the 2014 UK local elections.

The factors that can explain variation in political engagement include:

- Language barriers, especially among recently arrived immigrants.
- Lack of trust in politicians, strongest among the young and some ethnic minority groups.
- Feeling that you have no influence, strongest among minorities.
- Lack of belonging to a community giving a feeling of isolation.

This is important because the very communities that need regeneration the most are often the least engaged in the political processes that influence regeneration.

The degree of **place attachment** is important in explaining engagement. It varies according to age, ethnicity, the length of time people have been residents in an area and by level of deprivation. A famous example of this is Coin Street in London. This is an area on the South Bank of the Thames, close to the London Eye. In the mid-1970s plans were made to redevelop what was by then a run-down, deprived, deindustrialised area. However, local residents strongly objected to the commercial development plans and launched a campaign to save the area, eventually buying the land themselves and setting up a community group, Coin Street Community Builders (CSCB). In 2016 CSCB still own and manage social, affordable housing in the area.

However, Coin Street was an established (if deprived) community. People are likely to feel much less attached to places if they have recently arrived (immigrants), are temporary residents (students) or are young and can't see a future for themselves in a place.

**Sink estates** are council housing estates that are the least desirable to live in and have the shortest waiting lists for housing. They tend to house the lowest income, most in need residents.

**Election turnout** is the percentage of people who have a right to vote in an election that actually do vote.

## Knowledge check 23

Which has the higher voter turnout, local or national (general) elections in the UK?

## Exam tip

Find out what the biggest local and general election turnouts have been in the areas you have studied.

**Place attachment** is the bond between an individual or community and a location. It might be thought of as how much people care for a place.

Lower Falinge is a 'sink estate' in Rochdale housing about 1000 people, which was built in the 1970s. It has consistently been ranked one of England's most deprived areas:

■ In 2013, 72% of people living there were unemployed.
■ Around 4 out of 5 children on the estate lived in poverty.

You might think that there would be universal agreement that the estate should be bulldozed and replaced with something better, i.e. regenerated. In fact, the quotes from local people and residents in Table 17 suggest a much more complex relationship with the estate.

**Table 17** Views about the Lower Falinge estate, Rochdale

| | |
|---|---|
| 'It's nice. I've been here two years. Nice people. No problems. I came here because I didn't know anyone and now I'm interacting with people — just another member of the community.' **32-year-old asylum seeker** | 'There's more foreign people in the area but some of them are nicer than the people who were in the area before. It's no different to anywhere else though. You have nice people and bad people everywhere.' **Shop owner** |
| 'It's good if we demolish and start again and have a mixture of families there — I think it'll improve the area, for all the negativity and the crime rate, I think it'll be good.' **Local councillor** | 'No [it should not be knocked down]. And people are sick of reading negative things about their community.' **Chairperson of the Tenants Association** |
| 'Where would everyone go? It's a good place to live, it's just outdated.' **Resident** | 'They need CCTV, it's a disgrace. It needs demolishing.' **Resident** |

Source: *Guardian* and *Manchester Evening News*

There is evidence of conflicting views over what to do with the Lower Falinge estate. Despite the estate's serious problems, it is clearly 'home' to many people and they are attached to it. Even recent arrivals like asylum seekers seem to have developed an attachment. Others seem to feel a fresh start is needed. Lower Falinge could be regenerated without demolition which might retain the community in the area, although that option, i.e. regenerating run-down flats, is often much more expensive than demolition and rebuilding.

## Evaluating the need for regeneration

Deciding whether areas need regeneration requires the use of a range of different evidence. Some of this will be statistical data. These include:

■ census data about population growth and decline, age categories, ethnicity and health
■ IMD data which specifically identifies, at a small spatial scale, areas that are deprived and breaks this down further into seven deprivation domains
■ labour force surveys which tell us what average incomes in an area are, the types of jobs people do and whether they work full or part time.

These **quantitative data** are very useful as they provide a measure of the extent of social, economic and environmental problems in an area and can be used to compare places.

**Knowledge check 24**

Is Lower Falinge estate in Rochdale a sink estate or a gated community?

**Exam tip**

Watching television and YouTube clips about places is useful, but make sure your exam answers refer to a range of evidence not just media representations of places.

**Quantitative data** are numerical data that have set values, i.e. they measure the quantity of something.

Other media, such as images, television documentaries, blogs and even art works can give a sense of a place. Some of these are controversial. Television is a very powerful form of media. Documentaries such as Channel 4's *Benefits Street* (set in Birmingham and Stockton-on-Tees) can illustrate the need for communities to be regenerated but also provoke strong reactions that question whether some places 'deserve' investment. Similarly, Channel 5's *Benefits by the Sea* and *Benefits Britain* were both in part set in Jaywick in Essex. This is one of England's most deprived places by any measure. It could be argued that decisions about whether or not to regenerate areas like Jaywick should be taken on the basis of need, i.e. be data-driven, rather than be influenced by selective media representations that risk stigmatising people who live in deprived locations.

# How is regeneration managed?

- The UK Government is responsible for planning regulation and regeneration polices, as well as major infrastructure investment and immigration, which provides a national context for local regeneration.
- Local councils are key players in the success of local economies, and delivering regeneration in urban and rural places.
- Rebranding is an important element of regeneration, as it attempts to change people's perception of places, both urban and rural.

## UK government policy and regeneration

Regeneration is essentially a local process because only small areas are regenerated. However, much of the funding for regeneration comes from national government (and the EU) in the form of grants. As of 2016–17 the future of EU regeneration funding is unclear, because of the UK's decision to exit the EU. Local regeneration needs to be understood in the context of national government policy. This has tended to shift between regional and local strategies as shown in Table 18.

**Table 18** Regeneration policy since 1950

| 1950–1980 | Regional Policy & New Towns | Grants were directed at depressed regions, supported by new road development and the movement of government jobs out of London into the regions. New towns were built as inner city slums were cleared. |
|---|---|---|
| 1980–2000 | Urban Development Corporations (UDCs) and Single Regeneration Budget (SRB) | Focus on inner-city regeneration in deindustrialised areas in northern cities and London, led by quangos which were free from local council control and many normal planning regulations. |
| 1998–2010 | Regional Development Agencies (RDAs) | Regeneration was led by RDAs which decided how to spend government grants within large regions, like the North East and South West. |
| Post-2010 | Local Enterprise Partnerships (LEPs) | A much more local policy, focused on regeneration and job creation in specific small areas. There were about 40 LEPs in England in 2015. |

A **quango** (quasi-autonomous non-governmental organisation) is an organisation given the power to do a task that might be expected to be done by government. They are government funded, but act independently.

A feature of the UK since the 1950s has been a distinct **North–South divide**. Regeneration has attempted to reduce this divide. National infrastructure investment has also tried to reduce the divide by improving transport accessibility in order to improve economic growth:

- The UK motorway network, begun in the 1950s, has attempted to reduce road travel times between regions.
- The proposed High Speed Rail network between London, Birmingham, Liverpool, Manchester and Leeds is a planned attempt to connect the poorer regions to the economic core in London and the South East.
- Since the 1960s, there has been considerable investment in regional airports like Newcastle, Manchester and Glasgow.

Local regeneration takes place within the context of these national developments.

The UK has a number of **planning policies** that are important in the context of regeneration:

- Greenbelt land, which surrounds most large cities, cannot be built on; it is protected greenspace, usually farmland.
- Conservation areas like National Parks have strict planning regulations that limit the development of all but small-scale residential and commercial schemes.
- Planning permission is often dependent on a scheme including 'planning gain'; in other words a scheme for new private homes might be given permission if it includes a certain percentage of **affordable homes** or improvements to existing roads or parks, paid for by the developer.
- Planning laws allow for some developments that are 'in the national interest' such as fracking for shale gas, or quarrying, or new road building, even when these would not normally be allowed.
- National house-building targets set at 240,000 new homes per year in 2007 and revised to 200,000 per year for 2010–15 (these targets have never been met).

A very important context for understanding regeneration is the UK's shortage of homes and very high house prices. Demand for housing is very high in London, the South and East of England. This means house prices here are high — and in many locations unaffordable for most people (Figure 16).

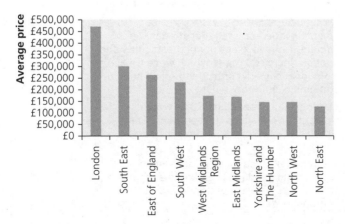

**Figure 16** Average house prices by region in England 2016

> The **North–South divide** is the difference in wealth between the industrial North, Midlands and North West and the more prosperous South and East of the UK.

> **Knowledge check 26**
>
> Has regeneration policy since 2010 in the UK been focused on local areas or regions?

> **Planning policies** are laws that regulate what can be built and where. They aim to prevent random, unplanned development that would harm the natural environment and create transport chaos.

> **Affordable homes** are those, usually for rent, which can be afforded by families on low incomes.

> **Exam tip**
>
> Make sure you are clear about the different roles of national government and local government in the planning process.

The housing shortage in the UK means:

■ There was a shortage of about 500,000 homes by 2016.

■ 240,000 homes need to be built each year to meet current demand.

■ For the last ten years, only 100,000–150,000 new homes have been built each year.

However, planning laws like greenbelt policy make releasing new land for housing very hard to achieve. In addition, most demand is actually in the South and East, the areas with the least capacity to find new land for housing. Lack of ability to build new housing can prevent economic development and regeneration by limiting the opportunities for new people to move into an area.

It could be argued that other government policies have contributed to housing shortages:

■ Immigration: large-scale immigration from the EU, especially since 2004, contributed to increasing the UK population from 59 million in 2001 to 63 million by 2011. An open-door immigration policy may have economic benefits but it also creates demand for new homes.

■ Deregulation: the UK is very open to foreign investment, including foreign people and companies buying property. In 2016 the *Guardian* estimated that 40,000 London properties were owned by offshore tax havens. These properties may not be lived in, or even rented.

■ Second homes and holiday homes: there are few restrictions in the UK on people buying houses to rent out, e.g. as a holiday let or buying a second home. In some rural areas a large percentage of houses may not be available to local people.

All of the factors above tend to reduce housing supply and increase prices of other homes. Longer term, there is a risk that very high house prices will prevent investment in some locations. UK-based or foreign investors could be put off investing in areas where housing for workers would be very expensive.

## Local government policy and regeneration

Most planning decisions are made by local councils at local level. Councils draw up plans, called Unitary Development Plans, which identify:

■ areas for new housing

■ priority areas for regeneration

■ new roads and other major infrastructure

■ areas for commercial development, i.e. factories, offices and retail.

Such plans aim to create environments that are attractive to both people and business (UK-based and foreign investors) and therefore create a successful place. This means planning has to provide a range of spaces such as:

■ retail parks and shopping centres

■ business parks for office functions, and industrial parks for manufacturing and distribution.

Increasingly a key goal is to attract high-value quaternary industry in fields such as ICT, pharmaceuticals and biotechnology, nanotechnology, 3D printing and space research.

Science parks are a key planning mechanism to deliver this. There are over 100 science parks and business incubators (smaller sites, for start-up companies) in the

**Knowledge check 27**

From Figure 16, what was the average London house price in 2016?

...................................

**Science parks** are industrial and business parks focused on the quaternary industry and usually involve at least one university as a key partner.

UK, owned by TNCs, universities and local councils. Perhaps not surprisingly most are in London, the South and South East.

NETPark (North East Technology Park) is a science park in Sedgefield, County Durham which is partnered by Durham University. Development, on a former hospital site, began in 2000 and the park now hosts 25 companies employing about 400 people. Other partners include Durham County Council, the UK Space Agency and Business Durham — the economic development agency for County Durham. Business Durham 'manages a portfolio of business property and excels in finding the right space for businesses to grow — commercial office space, modern laboratories and industrial property units' — in other words, its job is to attract domestic and foreign investment to County Durham.

Regeneration in the twenty-first century is about partnerships between the public and private sectors. Often public money from taxes is used as a **pump-priming** mechanism to 'lever in' private investment. In order for partnerships for regeneration to work they need to include as many **players** as possible:

- Chambers of Commerce: these organisations represent business and industry in a local area, and persuade local and national government to invest in infrastructure, education and skills training.
- Trade unions: their job is to represent workers in particular industries over issues like pay and conditions; their support is important to many companies that want good worker-manager relations.
- Education: schools, further and higher education establishments provide the skilled workers modern businesses need so they are key partners.
- Local groups: regeneration and development involves change, so local environmental groups, historical preservation societies and even wildlife groups need to be involved so that change does not lead to negative local consequences.

Regeneration plans can lead to conflict between different players. Inevitably regeneration leads to change:

- in the physical fabric of an area in terms of buildings, street patterns and possibly historic landmark buildings
- in the population of places, as newcomers move into regenerated areas.

These changes should be positive, but they are not always viewed in that way. We saw how in the 1970s the residents of Coin Street resisted attempts to develop commercial, hotel and retail space in their area (page 35).

- In the 1980s the regeneration of London's Docklands by the London Docklands Development Corporation (a UDC) sparked protests by existing residents who felt the service sector jobs being created, and expensive apartment housing, was doing nothing to reduce poverty in the area.
- Around 25 years later residents in Newham, the site of the London 2012 Olympics, also felt that in some cases their needs were being ignored as the huge regeneration of the Olympic Park took place (Figure 17), e.g. protests by residents of the Clay Lane Housing Cooperative which was **compulsory purchased** in 2007 to make way for Olympic site developments.

**Exam tip**

You need a named example of a science park to use in the exam.

**Pump-priming** means using money from national and local government to make an area more attractive to investors by improving derelict sites, transport, power and water supply, so that private companies choose to invest.

**Players** are the decision makers and other groups who have an interest in a particular plan or issue; the people affected by the changing geography of a place.

**Knowledge check 28**

What type of industry and jobs do science parks attempt to attract?

**Compulsory purchase** occurs when existing homes or businesses must be demolished to make way for new developments. Compensation is given, but the purchase cannot be prevented.

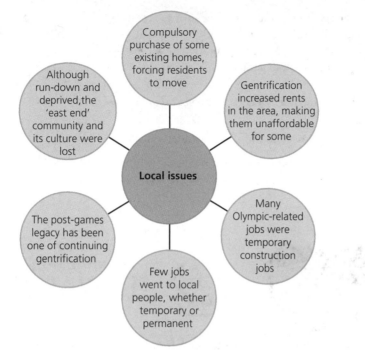

**Figure 17** Local issues surrounding the 2012 London Olympics regeneration

# Regeneration and rebranding strategies

Regeneration often focuses more on economic sectors based on the existing strengths of an area (such as an attractive physical landscape, good transport access, history and existing buildings or demand for new services). Some examples are shown in Table 19.

**Table 19** Regeneration strategies

| Urban areas | Rural areas |
|---|---|
| **Retail**<br>Major shopping malls, e.g. Westfield in Stratford or Meadowhall in Sheffield, which are 'destinations' as much as places to shop. | **Media themes**<br>Tourist trails based on popular TV programmes such as *Heartbeat* in the North York Moors or *Last of The Summer Wine* in Holmfirth, West Yorkshire. |
| **Heritage tourism**<br>Historic sites and attractions such as Hartlepool's Maritime Experience or the Titanic Quarter in Belfast. | **Outdoor pursuits**<br>Walking, mountain biking and climbing in Galloway Forest Park in southern Scotland or Zip World in Blaenau Ffestiniog's old slate quarries in Wales. |
| **Sport & Leisure**<br>Regeneration linked to major sporting events such as the London 2012 Olympics in Stratford, or the 2002 Commonwealth Games in East Manchester. | **Farm diversification**<br>Many private farms have shifted their focus from food production to camping, organic foods, shops and holiday cottages often with the support of public money, e.g. EU grants. |
| **Arts & Culture**<br>Landmark cultural buildings such as the Baltic art gallery and Sage music venue at Gateshead Quayside, or MediaCity in Salford. | **Sustainable rural livelihoods**<br>Regeneration focused on renewable energy and natural resources, such as Kielder Forest in Northumberland providing people with a new source of income. |

**Landmark**, or flagship, buildings are major new or redeveloped buildings with high visual impact designed to stimulate further regeneration.

**Exam tip**

A range of examples of different approaches to rebranding is needed, rather than one big case study.

**Knowledge check 29**

What is heritage rebranding based on?

Regeneration usually involves **rebranding**. This is important because successfully regenerated places become locations people want to live in, work in and visit as well as making them more attractive to investors. Most regeneration is not just for the existing population but attempts to draw in new people. This is done by using different types of media:

- Positive news stories and public relations during regeneration to get an area known.
- Advertising in newspapers and online.
- The use of logos and slogans to project an image to the outside world.

Logos and slogans have become especially important in a media-obsessed world. Places need to catch the attention of potential visitors and investors quickly. Some examples are shown in Figure 18.

> **Rebranding** means changing the perceived image of a place to outsiders, as well as physically regenerating an area.

| | | |
|---|---|---|
| Malton in Yorkshire is marketed as a food town. | Rural Scotland markets its landscape and wildlife. | Hartlepool stresses its maritime heritage as a historic port. |

**Figure 18** Rural and urban destination logos

## Rebranding deindustrialised places

Deindustrialised UK cities have proved hard to rebrand. Almost by definition the closure of industry, loss of jobs and spread of derelict land is not the stuff of marketing dreams. Places like Liverpool, Newcastle and Sheffield have had to be creative in order to produce a new, more attractive image. Usually this has involved:

- Turning their industrial history into a heritage asset, with museums, historic trails and public art works celebrating past achievements.
- Redeveloping warehouses and old industrial buildings into apartments, shops, restaurants and office space.
- Building modern apartments and hotels on land once occupied by industry.
- Using local art, artists and music to attract visitors.

In many of these cities canals, river-frontages and quaysides have been regenerated and turned into marinas and canals for leisure.

Liverpool is a good example of a deindustrialised UK city that has regenerated and rebranded in a number of ways as shown in Table 20. Liverpool has attempted to become a 'destination' for leisure, tourism, arts and culture. In 2014 Liverpool was the sixth most visited UK city by international visitors, showing that The Beatles' heritage in particular has an international dimension.

**Table 20** Regenerating and rebranding Liverpool

| Heritage | Between 1981 and 1988 Merseyside Development Corporation (a UDC) regenerated 320 hectares of the derelict but historic Albert Docks into a maritime museum, shops and apartments.<br>In 2004 the historic docks and Mersey waterfront became a UNESCO World Heritage Site. |
|---|---|
| Culture & Arts | The Tate Liverpool art gallery opened in the Albert Docks in 1988 followed by 'The Beatles Story' museum in 1990.<br>In 2002 Liverpool Speke Airport was rebranded as Liverpool John Lennon Airport, stressing the connection with The Beatles.<br>In 2008 Liverpool was European Capital of Culture, leading to an investment of about £4 billion in arts and infrastructure. |
| Retail | In 2008 the city centre was regenerated as Liverpool ONE, a £900 million shopping and leisure hub. |
| Living | Princes Dock has been regenerated and includes Liverpool Cruise Terminal (2007), Malmaison and Crowne Plaza hotels as well as apartments within a £5.5 billion regeneration called Liverpool Waters. |
| Technology | Liverpool Knowledge Quarter is a quaternary sector regeneration cluster including Science Park (2006) and Life Science Central (2013) partnered with Liverpool John Moores University. |

# Rebranding rural places

Rural areas in the **post-production countryside** have a tougher time rebranding compared to urban areas because of their relative isolation. It is very hard to attract visitors and investors to northwest Scotland or north Northumberland. Rural areas frequently focus on quite a narrow 'brand' to attract a particular type of visitor to spend their money.

It is important to get visitors to stay in an area for at least one night, as they then spend money on accommodation and food. In order to achieve this:

- There needs to be a range of accommodation options from campsites to bothies, B&Bs and expensive hotels.
- There needs to be plenty to do and see.

The exception to this is in **accessible rural areas** close to cities. Farms in these places find **diversification** easy because city people will drive out into the countryside to visit a tea room, farm shop, go paint-balling, etc., on a regular basis (Figure 19).

Which famous 1960s band has been used to help rebrand Liverpool?

The **post-production countryside** refers to rural areas that no longer make most of their income from food production and other primary sector employment like fishing, quarrying and forestry.

**Accessible rural areas** are those within 15–30 minutes' driving time of cities. They are often home to commuters as well as farms.

**Diversification** means farms, and rural areas generally, finding new income to replace farming as the main source of income, especially in leisure and tourism.

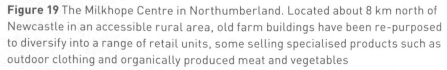

**Figure 19** The Milkhope Centre in Northumberland. Located about 8 km north of Newcastle in an accessible rural area, old farm buildings have been re-purposed to diversify into a range of retail units, some selling specialised products such as outdoor clothing and organically produced meat and vegetables

In more remote rural areas the strategy is to attract people and get them to stay. Examples include:

- **Literary associations**

  Hardy Country: novelist and poet Thomas Hardy has been used to rebrand the rolling hills of Dorset around Dorchester, Bockhampton and Stinsford. Places associated with Hardy's life and books form a route around the region encouraging people to stay and spend.

- **Heritage**

  Northumberland Coastal Route: stretching from Alnmouth to Berwick and including a Heritage Coast, this area (once important for fishing and coal mining) promotes its coastal castles (Warkworth, Dunstanburgh and Bamburgh) as well as the historic Lindisfarne Gospels on Holy Island. Nearby Alnwick Castle was the set for some of the Harry Potter films, attracting younger people.

- **Outdoor adventure**

  Blaenau Ffestiniog in Wales was once an important centre for slate production. Now the abandoned slate quarries and hills have become a centre for adventure tourism including Zip World (zip wire rides) and Bounce Below (trampolining in caves), and the Antur Stiniog downhill mountain bike trails.

Rural regeneration can be very successful, but it is worth noting that at least in the UK the rural economy tends to be very seasonal and weather dependent. In addition some outdoor activities such as walking and going on a 'Sunday afternoon drive' do not bring much money into the rural economy. Rural areas have to fight hard to attract people and get them to stay and spend.

# How successful is regeneration?

- Economic, demographic, social and environmental indicators can be used to measure the success of regeneration.
- Different groups of people judge the success of urban regeneration in different ways.
- Different groups of people judge the success of rural regeneration in different ways.

## Measuring the success of regeneration

Regenerated areas should shows signs of improved economic performance and quality of life if regeneration has been successful. A key indicator of success is population change. Demographic growth indicates that areas are popular and people are moving in.

Liverpool is a case in point. Its population declined from 683,000 in 1961 to 439,000 in 2001. However, the 2011 Census showed an increase to 466,000 — the first growth in 50 years. This alone suggests Liverpool's long period of regeneration since 1980 has finally paid off.

> **Exam tip**
>
> Make sure you know if your rural rebranding examples are accessible or remote, as rebranding differs between the two types of area.

> **Demographic** is the technical term for population total numbers and structure (age groups and males versus females balance).

Figure 20 shows the population of an area on the River Tyne quayside in Newcastle (called Ouseburn) which has been regenerated over the last 20 years. Between 2001 and 2011:

- The working population aged 25–64 increased, suggesting that people are moving into new apartments and houses.
- The 70+ population declined, which could indicate that older residents are being pushed out by regeneration.
- There was a huge increase in the 15–24-year-old population, because much of the most recent regeneration has involved new student accommodation, i.e. **studentification**.

Areas populated with students are not necessarily welcomed by all, due to noise issues, the transient nature of the population and the fact that students tend to be up all night!

**Studentification** happens when an area becomes popular with students, and it can lead to dramatic changes in the age profile of an area as student accommodation increases.

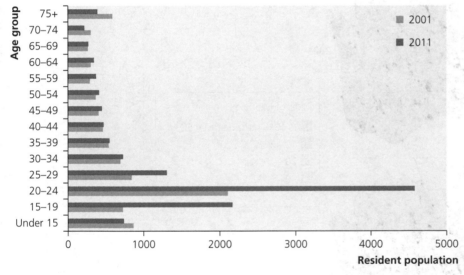

**Figure 20** The population structure of Ouseburn, Newcastle in 2001 and 2011

**Exam tip**

Revising some demographic data is important as they are a key indicator of regeneration success.

**Knowledge check 31**

What is the name for the process when an area's population changes as a result of college and university students moving in?

Newcastle's quayside regeneration can be assessed in other ways. Figure 21 shows deprivation levels from the IMD in 2004 and 2015. Notice that:

- Areas along the quayside have improved, moving out of the 10% most deprived and 10–20% most deprived categories and into the 20–50% range.
- Areas to the east and west remain in the 10% most deprived category.

These data suggest successful regeneration, but also suggest greater inequality between the now regenerated areas and areas close by that have seen no regeneration.

Liverpool has made similar progress. After the city was European Capital of Culture in 2008, 85% of residents felt the city was a better place to live. Merseyside Development Corporation created 22,000 jobs between 1981 and 1998 and attracted £700 million of private investment. Between 1998 and 2008 a further 25,000 jobs were created.

However, although Newcastle and Liverpool have changes in numbers employed and levels of deprivation, relatively they are still worse than other places. Child poverty in Liverpool fell from 34.7% of children in 2006 to 32.5% in 2011, but the England average in 2011 was 20.1% and in Reading it was 17.8%.

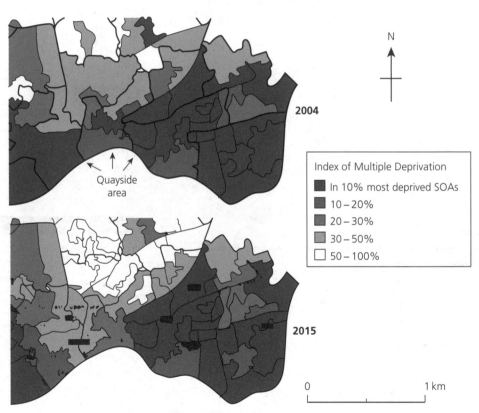

**Figure 21** Deprivation levels in Newcastle in 2004 and 2015

## Success and social progress

A key question is whether regeneration actually reduces the inequalities it sets out to. A report in 2013 by the Work Foundation looked at inequality in UK cities based on differences in wages:

- The top five most unequal cities were London, Reading & Bracknell, Portsmouth, Guildford and Aberdeen.
- The top five most equal cities were Bradford, Plymouth, Barnsley, Stoke and Burnley.

Overall, northern deindustrialised cities were much more equal than their more successful southern counterparts, because most people are poor. Greater Manchester is one of the northern industrial cities that has seen waves of regeneration since the 1980s in Salford Quays, East Manchester for the 2002 Commonwealth Games and in the city centre after the IRA bombing in 1996. **Social progress** data for three areas within Greater Manchester are shown in Table 21 for 2001 and 2011.

**Table 21** Social progress data in Greater Manchester

| Area | Year | Working age, employed full-time (%) | Working age, no qualifications (%) | Infant mortality (per 1000 live births) |
|---|---|---|---|---|
| Salford | 2001 | 39.3 | 35.5 | 5.9 |
| | 2011 | 39.2 (−0.1%) | 27.1 (−8.4%) | 5.7 (−0.2) |
| Manchester | 2001 | 32.9 | 33.9 | 9.3 |
| | 2011 | 33.7 (+0. 8%) | 23.1 (−10.8%) | 5.6 (−3.7) |
| Trafford | 2001 | 43.4 | 24.7 | 5.6 |
| | 2011 | 41.9 (−1.5%) | 18.6 (−6.1) | 3.5 (−2.1) |

Notice that the data in Table 21 show a number of changes:

■ All three areas have improved overall, especially improved health measured by falling infant mortality and better education measured by the percentage with no qualifications.

■ Manchester is the only area with more full-time employees and it has the best improvements in qualifications and infant mortality.

■ Salford has experienced the least progress.

There is some evidence here that Manchester has made most progress relative to the other two areas. This is confirmed by the 22% increase in Manchester's population 2001–14 compared to only 11% in Salford and Trafford.

# Improving living environments

Regeneration often attempts to improve environmental quality by:

■ Redeveloping derelict land and buildings, and removing **contaminants** from former industrial sites.

■ Creating pedestrianised zones, introducing traffic calming and making streets more friendly for people.

■ Creating new parks and green spaces, planting trees and creating lakes and wetlands.

■ Putting in place new street furniture, pavements and lighting to improve the design of areas.

■ Regenerating housing with double-glazing and insulation to reduce indoor noise, damp problems and lower energy costs.

These environmental improvements have a number of aims:

1 To reduce air pollution levels; this in turn improves people's health.

2 To create spaces for people to walk, play and play sport, which in turn could lead to healthier, more active lifestyles.

3 To make urban environments safer, especially for children, pedestrians and cyclists.

4 To improve living conditions so people live in warm, dry, secure homes.

**Social progress** means how a community improves its quality of life, health and welfare over time.

**Exam tip**

It is important to be able to compare areas in the exam, in order to make judgements about how successful regeneration has been.

**Contamination** refers to chemical and dangerous organic substances in the ground, which are a legacy of previous industrial use.

The Index of Multiple Deprivation (IMD) includes both indoor and outdoor environmental quality as part of the Living Environment Deprivation domain.

The 2012 London Olympics in Stratford, east London is a good example of environmental improvement, much of which occurred during the construction phase across an area of 350 hectares of abandoned and derelict land (Figure 22) and is less 'obvious' than the new sports facilities:

- 100 hectares of new greenspace was created with 4000 trees, after 600,000 tonnes of soil had been cleaned of contaminants like arsenic, bitumen and ammonia.
- 3 km of rivers and canals were cleaned and in many cases, replanted.
- 230,000 cubic metres of contaminated groundwater was removed and cleaned.

Most of the accommodation for athletes eventually became new homes for 2800 people, who live in a much cleaner environment than before the 2012 Olympics. In addition improvements in rail, bus routes and cycle routes improved transport in the area, potentially reducing air pollution.

Figure 22 The 2012 Olympic Stadium during construction in 2010

## Urban and rural stakeholders

A key question to ask about urban and rural regeneration is who has it benefited? Different **stakeholders** have very different views of this because they use different **criteria** to judge success. Some regeneration schemes have a questionable impact on local people. A good example is Salford Quays in Greater Manchester. Once a thriving industrial inland port, by the 1980s the place was derelict and abandoned. Table 22 gives a brief history of its regeneration.

**Knowledge check 32**

What aspect of people's wellbeing does outdoor air pollution affect in a negative way?

**Exam tip**

It is worth researching recent regeneration schemes, like the 2012 Olympic site, to see if any further changes have taken place.

**Knowledge check 33**

Why is former industrial land often not able to be used for regeneration without being cleaned up first?

**Stakeholders** are any groups or individuals involved in, or interested in, regeneration. They range from residents, environmentalists and businesses to local councils and planners.

**Criteria** are the standards or measures people use to judge whether or not something is a success.

**Table 22** Regenerating Salford's Docklands

| Salford Quays Development Plan | Landmark buildings | MediaCityUK |
|---|---|---|
| 1985–1990: about 90 hectares of former industrial land were developed by Salford City Council and private investors. | 2000: The Lowry theatre and gallery opened. 2001: The Imperial War Museum north opened. | 2007–2011: developed by the property company Peel Holdings and housing the BBC and ITV Granada among other media businesses. |

Salford Docklands has been transformed, but for whom? In 2016 apartments in NV Buildings, a development built in 2004, were on sale for up to £825,000 — hardly for local people. Other aspects of Salford's regeneration can also be questioned.

- In one recent Salford Quays apartment development, called the Dock Office, 50% were sold to local people but 25% to Chinese investors and 25% to UK residents living overseas.
- In 2012 it was reported that only 24 of the 680 new jobs at the BBC in Salford had gone to local people.
- In 2013 local historians and local people were upset when two iconic industrial quayside cranes were demolished — one of the last icons of Salford's past importance as an inland port.
- 'Salford Quays' as the area is now called is actually a made-up name — the historical name is Salford Docklands — but this sounded too industrial when regeneration began in the 1980s.

It is possible to see Salford Quays from many different viewpoints as shown in Table 23, because different stakeholders have different desired outcomes from regeneration.

**Table 23** The desired outcomes of regeneration

| Stakeholder | Desired outcomes | Measured by: |
|---|---|---|
| Local council | External image is important to attract investment, so landmark buildings and interesting architecture are important, but so are reduced deprivation levels. | ■ Job creation numbers ■ Area of vacant/derelict land brought back into use ■ IMD data trends |
| Existing residents | Better housing, community facilities and job opportunities plus an improved environment. Hard-to-measure factors such as 'community spirit' may be important. | ■ Rising incomes ■ Improved health and life expectancy ■ Increased access to services |
| Property developers | Motivated by profit, so will maximise sales values and rental values. Image is important to drive sales. | ■ Profit versus investment ratios ■ Number of investors ■ Increase in land value |
| Local businesses | Increased local population, especially of wealthier residents to boost trade and profits. | ■ Rising population, especially the young ■ Hiring rates of new employees |
| National government | Regeneration that fits in with national priorities such as the Northern Powerhouse, plus reduced dependency on benefits. | ■ Reduced out-migration ■ Increases in regional output / GVA |

The '**Northern Powerhouse**' was a government policy to increase the economic power and significance of northern cities, especially Greater Manchester.

It is worth considering the occasions when regeneration is not successful. Examples are rare, because most regeneration is carefully planned. Doncaster's Earth Centre is an example. Opened in 1999, it was built on the 160-hectare former coal mine site of Cadeby Main Colliery near Denaby. The derelict land was reclaimed and an attraction focused on sustainable lifestyles as part of a 'green theme park' was built with a grant of £42 million from the Millennium Commission (National Lottery money). It closed in 2004 due to low visitor numbers. Around 75 employees lost their jobs because it attracted only 37,000 of the 150,000 visitors needed per year to make it viable. There are a number of possible reasons for this failure:

- The location was not very accessible, and the area is not well known for tourism.
- Local people's **lived experience** of the area was coal mining and its heritage, not ecological issues.
- The reality of South Yorkshire is of an industrial and agricultural place, not an 'environmental' or 'ecological' place.
- Interpretive centres tend to lack the attractions (rides, etc.) that many families want.

Rural regeneration can also be contested. Balmedie is a village in Scotland, on the coast just north of Aberdeen. Between 2003 and 2012 its population grew from 1850 to 2520 people, an increase of 36%. This is a traditional farming area, with some people employed in Aberdeen's oil industry and many people commuting to Aberdeen for work.

In 2006 Donald Trump, a US businessman and politician elected president of the USA in 2016, bought the Menie Estate in Balmedie. Trump planned to invest £1 billion creating a luxury golf resort (Trump International Golf Links) employing 1200 people and including a luxury hotel and 1500 new houses. However, opposition to the plans was extensive despite the promised jobs (Table 24).

**Table 24** Opposition to Trump International Golf Links

| | | |
|---|---|---|
| Local groups such as 'Tripping up Trump' were set up to oppose the plans which were seen as too large and intrusive. | Scottish Natural Heritage opposed the destruction of sand dunes, some of which are part of an **SSSI**. | Michael Forbes, a local farmer, has refused to sell his land to Trump despite being offered £450,000. |
| The local council opposed the development as some of it encroached on greenbelt land. | RSPB Scotland opposed the destruction of habitats for birds. | Residents objected to the threat of their homes being bulldozed to make way for the development. |

In 2007 Aberdeenshire Council turned down Trump's planning application, but it was approved on appeal in 2008 by the Scottish Government. Trump fought a losing battle against plans to build an offshore windfarm opposite the Menie Estate, eventually losing an appeal to the Supreme Court.

By 2016, Trump International Golf Links employed 95 people and some parts of the development have been completed but the golf course is not open in the winter.

This example shows that there are many different criteria by which to judge for successful rural regeneration:

- At a local scale regeneration can be opposed on environmental and community impact grounds.

**Knowledge check 34**

What industry was Salford Quays famous for before deindustrialisation and regeneration?

**Lived experience** refers to people's feelings about a place, what is important to them and what a place means to them. This is strongly related to a place's traditions, its community spirit and its history.

**Exam tip**

Make sure you revise examples of both contested rural and urban regeneration schemes.

An **SSSI** (Site of Special Scientific Interest) is a conservation area with particular importance in terms of wildlife.

**Knowledge check 35**

Which economic sector was Trump's plan in Aberdeenshire focused on: primary, secondary, tertiary or quaternary?

- Even in the Aberdeenshire region the promise of jobs, even in large numbers, failed to convince all that large-scale plans are a good idea.
- Nationally, i.e. within Scotland, the promise of jobs and investment seems to have swayed the Scottish Government.
- At UK level, targets for renewable energy, i.e. windfarms, seem to have 'out-trumped Trump'.

Both the locals' objections to Trump's plans, and Trump's objections to wind turbines are classic examples of **Nimbyism**.

**Nimbyism** (Not In My Backyard) refers to the fact that most people object to change in their own area (their own 'backyard') even if they agree with the principle behind the change.

## Summary

- Economic activity — how people earn a living and their prosperity — varies markedly from place to place, especially between urban and rural places.
- Places have very different functions and population characteristics in terms of age and ethnicity, which reflects past and more recent history and economic changes.
- Contrasting places like Reading and Middlesbrough have been affected by local, regional, national and global forces which helps explain their characteristics today.
- There is a major contrast between successful places like Santa Clara in California and deindustrialised places like Detroit, with very different experiences for the people who live there.
- Places have different levels of engagement and people in those places have different views on regeneration.
- Regeneration is influenced by national policy decisions about infrastructure and planning regulations, as well as migration policy.
- Local government has a key role in regeneration, especially in terms of attracting investment, which is one of the most important aspects of successful regeneration.
- In both rural and urban areas rebranding — to change the image of a place — has become increasingly important.
- Measuring the success of regeneration can be done by examining a wide range of data including on incomes, deprivation, population changes and environmental improvement.
- Urban areas such as Salford have been regenerated but different stakeholders have contrasting views on how successful this has been.
- Rural regeneration, such as in Aberdeenshire, can also be controversial.

# Diverse places

## How do population structures vary?

- Population structure varies from place to place and changes over time, as a result of a number of factors.
- The ethnicity, gender and migration characteristics of populations vary from place to place and change over time.
- Global, international, national and regional forces today and in the past have helped shape the populations of places.

This section of the guide compares two contrasting places in the UK: Southwark and Anglesey. You will have studied your own places. You should relate the places you have studied to the themes in the content guidance below. You could also use Southwark and/or Anglesey as additional case studies.

## Population structure

The UK's population, like that of most countries, is very unevenly distributed and changes at different rates depending on location. Figure 23 shows UK regional population change since 1981. The UK population grew from 56 to 63 million between 1981 and 2011, but not in all regions equally. Figure 23 suggests three distinct zones: no growth in the North, some growth in the middle and strong growth in the South. This pattern broadly reflects the economic prosperity of the regions, with the deindustrialised North losing out to the service sector economy of the South. It also reflects a persistent North–South divide in the UK.

The **North–South divide** is the difference in wealth between the industrial North, Midlands and North West and the more prosperous South and East of the UK.

**Knowledge check 36**

Which UK regions had population growth of 1 million people or more between 1981 and 2011?

| UK region | 1981 (million people) | 1991 (million people) | 2001 (million people) | 2011 (million people) |
|---|---|---|---|---|
| North East | 2.6 | 2.6 | 2.5 | 2.6 |
| North West | 6.9 | 6.8 | 6.8 | 6.9 |
| Yorkshire & The Humber | 4.9 | 4.9 | 5.0 | 5.3 |
| East Midlands | 3.8 | 4.0 | 4.2 | 4.5 |
| West Midlands | 5.2 | 5.2 | 5.3 | 5.5 |
| Eastern England | 4.9 | 5.1 | 5.5 | 5.8 |
| London | 6.8 | 6.8 | 7.3 | 7.8 |
| South East | 7.2 | 7.6 | 8.0 | 8.5 |
| South West | 4.4 | 4.7 | 4.9 | 5.5 |
| Wales | 2.8 | 2.9 | 2.9 | 3.0 |
| Scotland | 5.2 | 5.1 | 5.1 | 5.2 |
| Northern Ireland | 1.5 | 1.6 | 1.7 | 1.8 |

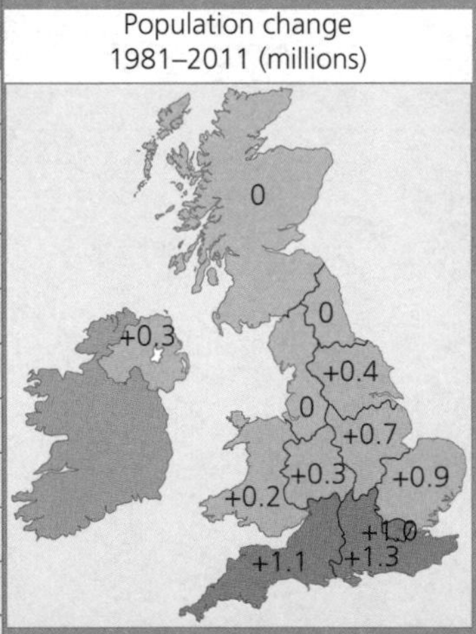

Figure 23 UK regional population change 1981–2011

Demographic structure is also highly variable between places. There is often a significant difference across the **rural–urban continuum** in both population structure and density as shown in Figure 24. Population has also changed in these areas, generally:

- Remote rural areas have experienced population decline, as have some inner cities.
- Suburban, rural–urban fringe and accessible rural areas have seen population growth.

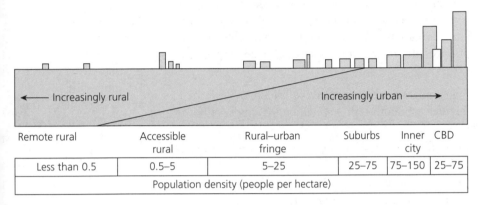

**Figure 24** The rural–urban continuum

As well as the number of people being different, population structure varies across the continuum. Table 25 shows that:

- Older people aged over 65 tend to live in rural areas.
- The percentage of over-65s in inner city areas is low.
- Suburban places have a high number of 0–15-year-olds.
- There is less variation in 0–15-year-olds than over-65s.

**Table 25** Young and old population for four places

| Area type | Example | % population aged 0–15 | % population aged 65 and over |
|---|---|---|---|
| Remote rural | Highland (Scotland) | 17.9 | 18.5 |
| Accessible rural | South Oxfordshire | 19.4 | 18.2 |
| Suburbs | Hillingdon (London) | 21.4 | 11.0 |
| Inner city | Southwark (London) | 18.5 | 8.0 |

Population differences can be explained by a number of factors:

- Urban areas are accessible by transport so have high populations, whereas rural areas have less good access so lower populations.
- Access matters for working-age people, but less so for the over-65s so retired people often live in rural areas that are more peaceful but also have fewer services.
- Remote rural areas are often mountainous and therefore access is even harder; they have limited transport connections and long journey times between places, so low population densities.

**Demographic** is the technical term for population total numbers and structure (age groups and males versus females balance).

The **rural–urban continuum** refers to the gradual transition from highly urban places with high population densities to remote rural places with low densities.

**Exam tip**

Make sure you know where your examples of places sit on the rural–urban continuum.

**Knowledge check 37**

Which type of area along the rural–urban continuum has the lowest population density?

- Upland areas have poor farmland suitable only for animal grazing, so even the farm population is few and far between.
- Suburban areas have high population densities partly as a result of history; after 1880 many middle-class people moved out of city centres into the suburbs and began to **commute** to work by rail — something that continues to this day.

To some extent, planning has contributed to the popularity of the suburbs, accessible rural areas and some parts of the rural–urban fringe. Since 1947 many cities have been ringed by a **greenbelt**. This has encouraged people to move out beyond the greenbelt and live in rural areas and commute to work in the city. In addition, greenbelts have put pressure on suburban areas to house people at higher densities.

Two places with very contrasting population structures are the Isle of Anglesey in North Wales and the London borough of Southwark. Population pyramids for both places are shown in Figure 25.

- Anglesey has an ageing population, as shown by its 'top heavy' pyramid with large numbers of people aged over 50.
- Southwark's population is dominated by people aged 20–40, i.e. working age, with few people aged over 65 but a large number of children under 10.

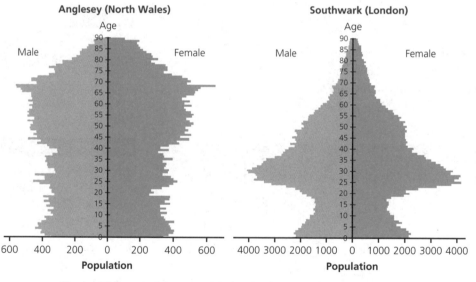

**Figure 25** Population pyramids for Anglesey and Southwark 2014
Source: ONS

Several factors can help explain the differences between the two populations:

- In 2014 there were 813 deaths in Anglesey, but only 615 births. This means **mortality** is higher than **fertility**, so population would fall if this continued.
- In Southwark the situation is different. There were 4600 births and only 1300 deaths in 2014, so population is growing.

Births and deaths are not the only factors that affect population structure. **Internal** and **international migration** can also have an impact:

**Commuting** is the daily journey to work by rail, car or bike.

**Greenbelts** are land surrounding cities that cannot be built on, usually farmland. Development sometimes 'leapfrogs' the greenbelt benefiting places just beyond the greenbelt edge.

**Exam tip**

It is useful to be able to sketch two contrasting population pyramids for the exam.

**Mortality** means death rate, the number of deaths per year usually measured as deaths per 1000 people.

**Fertility** is the average number of children born to women of childbearing age..

**Internal migration** is movement from one region to another within a country, whereas **international migration** is from one country to another.

- Inner-city Southwark attracted 5500 extra international migrants in 2014, which more than made up for the net loss of 2400 people who left Southwark for somewhere else in the UK.
- Rural Anglesey only attracted 60 extra international migrants in 2014, fewer than the 120 people it lost due to internal migration.

Anglesey's population declined by 0.3% in 2014, whereas Southwark's grew by 2.5%. Southwark's population is also much more **dynamic**. With a total population of just over 300,000 in 2014, Southwark had 36,000 new migrants arrive and 33,000 existing residents leave — a very high rate of population change.

## Population characteristics

In most of the UK the number of males and females is very similar, but not everywhere:

- Some cities have more males than females because some industries are dominated by male employees, such as the offshore oil industry based in Aberdeen.
- Rural areas often have more single men than single women, often explained by women being more willing to leave and move to an urban area to look for work.
- There are more male international economic migrants than females, so inner city areas often have more young male immigrants.
- Because women live on average longer than men, coastal retirement locations like Bournemouth have more older women.

**Ethnicity** is much more variable than gender. Table 26 shows this variability for Anglesey and Southwark. Anglesey is almost all white British, whereas Southwark has a very ethnically diverse population.

**Table 26** Ethnicity in 2011 for Anglesey and Southwark

| Ethnic group | Anglesey (%) | | Southwark (%) | |
| --- | --- | --- | --- | --- |
| White British | 96.6 | | 39.7 | |
| White Irish | 0.7 | 98.3 | 2.2 | 54.3 |
| White other | 1.0 | | 12.4 | |
| Asian Indian | 0.2 | | 2.0 | |
| Asian Pakistani | 0 | | 0.6 | |
| Asian Bangladeshi | 0.1 | 0.7 | 1.4 | 9.5 |
| Asian Chinese | 0.2 | | 2.8 | |
| Asian other | 0.2 | | 2.7 | |
| Black African | 0.1 | | 16.4 | |
| Black Caribbean | 0 | 0.1 | 6.2 | 26.8 |
| Black other | 0 | | 4.2 | |
| Arab | 0.2 | | 0.8 | |
| Other | 0.1 | 1.1 | 2.4 | 9.4 |
| Mixed/multiple | 0.8 | | 6.2 | |

As can be seen in Table 26 Southwark has **cultural diversity** whereas Anglesey does not. In general rural areas have low ethnic diversity, but they may have some cultural diversity. In areas like rural Lincolnshire there are large numbers of Polish, Lithuanian and Latvian workers (5–10% of the population). These economic migrants are ethnically white European but culturally they are different to the UK population. Many work on farms, in food processing factories and doing distribution jobs.

**Dynamic**, in relation to population, means changing rather than static. Change can result from high fertility rates and/or high levels of migration.

**Knowledge check 38**

In Southwark, which age groups make up the largest part of the total population?

**Ethnicity** refers to groups of people who share a common culture, ancestry, language and traditions — and often religion. Race (racial group) is based on physical/genetic characteristics.

**Cultural diversity** is a measure of how many different ethnic and cultural groups live in an area.

Accessible cities are more culturally diverse because of the availability of employment, whereas physically remote areas tend to have the least diversity.

In urban areas ethnic groups sometimes exhibit **clustering** in terms of where they live. Clustering is a form of segregation, meaning people who are different live separately from other people. In almost all places economic segregation takes place, i.e. wealthy people live in separate areas to poor people. However, ethnic segregation is a step beyond this.

Figure 26 shows clustering in Southwark, London. Certain wards have higher than expected concentrations of some ethnic groups. This happens for a number of reasons:

- Expensive riverside property in Surrey Docks has been bought up by wealthy European immigrants.
- The wealthier white British population tends to live in the southern wards, which are more suburban and furthest from the densely populated riverside.
- Lower income ethnic groups may be concentrated in areas with a large amount of local authority (council) housing.

**Clustering** refers to an uneven distribution of population in an area, so that people with similar characteristics are found clustered together in one place.

## Knowledge check 39

In Southwark, which is the largest ethnic group after white British?

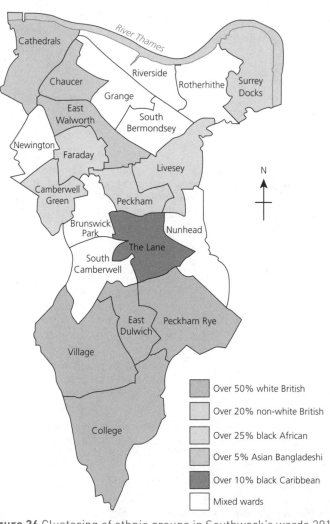

**Figure 26** Clustering of ethnic groups in Southwark's wards 2011

There are other reasons why people from ethnic groups tend to cluster, outlined in Table 27.

**Table 27** Internal and external explanations of ethnic clustering

| Internal explanations (actions and attitudes of the ethnic group) | External explanations (actions and attitudes of the rest of society) |
|---|---|
| New immigrants tend to live close to existing people from the same ethnic group, because they share a common language and experiences. | Estate agents or council housing officers may (consciously or unconsciously) help concentrate groups in particular areas. |
| Ethnically specific services — shops, places of worship, schools — encourage others to live nearby for convenience. | An existing population may leave an area if a new ethnic group begins to move in, making more housing available. |
| It may be felt there is 'safety in numbers' and stronger community ties if people live close together. | Prejudice in the jobs market prevents some ethnic groups gaining high enough incomes to live in some areas. |

Southwark is changing, both demographically and culturally because:

- Each year about 13% of its resident population changes because people move away, and new people move in.
- Southwark was 63% white in 2001, but 54% in 2011, so its cultural diversity is growing rapidly.
- Most of Southwark's population growth is in the 20–35 age group; in other words young, dynamic, educated workers.
- Population growth is driven by high fertility rates, because young immigrant populations have higher fertility than the population in general (and youthful populations have low mortality rates).

## Demographic and cultural change

Places, almost wherever they are, are influenced by regional, national, international and global forces today and in the past which affect them in both positive and negative ways. These forces are summarised for Anglesey and Southwark in Table 28.

**Exam tip**

When explaining ethnic clustering/segregation make sure you explain both internal and external factors.

**Identity** refers to people's feeling and perceptions, and their shared beliefs, traditions and ways of life.

**Table 28** Forces shaping Anglesey and Southwark

| | Anglesey | Southwark |
|---|---|---|
| **Global** | Globalisation allows Anglesey's seafood industry access to global markets, but in 2009 Anglesey Aluminium (owned by TNC Rio Tinto) shut down due in part to falling global demand; it employed over 500 people. | Global events like the 2012 London Olympics raised the profile of London as a world city, bringing tourism to Southwark's Southbank area. |
| **International** | Limited connections to the rest of Wales and the UK in part explain its ageing population, but there is a tourism industry. | As a London borough, it benefits from being within the EU's largest city and some global brands like BAE Systems and the *Financial Times* are located there. |
| **National** | As a location for economic activity, it is remote and peripheral although ferry connections to Ireland do benefit the area and its major port, Holyhead. | Cultural icons such as the Tate Modern art gallery, Shakespeare's Globe and National Theatre all give the borough a high cultural profile. |
| **Regional** | As an island off the Welsh coast, it has a strong identity of its own, as well as a strong Welsh identity (57% of the population are Welsh speakers, the highest in Wales). | Southwark is south of the River Thames, historically seen as the poorer relation to north London but also a bit more edgy, with more going on, and fewer tourists to deal with. |

Fairly or unfairly, all places have an image which they project and this shapes people's perceptions of the place as either positive or negative. This image can also have an effect on people in the place. Their identity may be affected if they perceive they are living in an area that has a positive or negative image. Figure 27 shows images of both places. They are very different. Southwark is a high density, bustling city area whereas Anglesey is remote rural. Older, retired and outdoor-type people might find the peace and seclusion of Anglesey very appealing, but be horrified by the congestion and noise of Southwark. On the other hand young people may perceive Anglesey as boring and isolated, whereas they see Southwark as teeming with social and economic opportunities. Students, other young workers and migrants are affected by these images and perceptions:

- Young people may feel they want to leave a place with a poor image.
- People are attracted to places with positive images.
- There are likely to be more job opportunities in places with attractive images because companies, like people, are attracted there.

**Exam tip**

Remember that the forces that shape places are caused by change at many scales, from local to global.

**Knowledge check 40**

What do 57% of people on Anglesey have in common?

**Figure 27** Images of Southwark (top) and Anglesey (bottom)

Since 2010, the UK Government has attempted to measure 'national wellbeing' by conducting a survey asking people how they feel about their lives. Results from 2015 for Southwark and Anglesey are shown in Table 29. The results are not dramatically different, but in all cases people in Southwark answered low/medium more than people in Anglesey. This might give an insight into how people perceive their place.

**Table 29** 2015 National wellbeing survey results

| How do you feel about: | Low/medium | | High/very high | |
|---|---|---|---|---|
| | Anglesey | Southwark | Anglesey | Southwark |
| Life satisfaction | 20% | 22% | 80% | 78% |
| Life is worthwhile | 16% | 19% | 84% | 81% |
| Happiness | 24% | 28% | 76% | 72% |

# How do different people view diverse living spaces?

- Urban places, and locations within towns and cities, are perceived differently by contrasting groups of people.
- Rural places are often perceived as 'idyllic' but not in all cases or by all people.
- Different types of evidence can be used to determine whether places have a positive or negative image.

## Perceptions of urban places

In the twenty-first century many Victorians viewed cities as dangerous and threatening, and places to avoid. At least this was the view of the middle and upper classes that did not have to work in factories and mills. This perception was caused by:

- Pollution from factories: during the Industrial Revolution this literally blackened the buildings.
- Poverty: the working class lived in inner city slum housing, with minimal sewerage and sanitary facilities.
- Congestion: many Victorian cities had worse traffic congestion than the same cities today.

There was also a perception of crime including pick-pocketing, petty theft and prostitution. The novels of Charles Dickens, such as *Oliver Twist*, only played on this perception. Wealthier Victorians reacted to this perception by:

- moving out to the **suburbs**, away from the **inner city**
- planning entirely new 'model' cities, such as Ebenezer Howard's garden cities of Welwyn and Letchworth.

The very same inner-city areas that the Victorians feared are today perceived as attractive places because of the range of economic opportunities and the variety of social and leisure activities found there. This is the case with Southwark in London.

- Borough Market attracts food lovers from all over the city.
- The area around the Tate Modern art gallery, opened in 2000, has become a trendy place to live.

**Suburbs** are areas of low-density, low-rise housing surrounding cities but attached to them so people can commute to work. Suburban houses typically have gardens.

**Inner-city areas** surround the Central Business District (CBD) of a city and in the UK they are often dominated by buildings dating from the Industrial Revolution, especially terraced housing which is high density with little greenspace.

**Knowledge check 41**

Where did many Victorian city-dwellers move in response to their dislike of industrial cities?

- The South Bank area around the OXO Tower, Gabriel's Wharf and the National Theatre has become a trendy area to eat and drink.
- Job opportunities abound in the nearby City of London and Westminster.

This is why Southwark attracts so many young workers, and internal and international migrants (see Figure 25 and Table 26).

Not all urban places are perceived in a positive way by all groups. Some urban places have a 'reputation', which is often based more on negative images from the past than the reality of today. Liverpool is an example:

- Riots in the Toxteth area of Liverpool in 1981 (and in 1985) were widely reported on TV.
- The 1982 TV series *Boys from the Blackstuff* portrayed the impact of deindustrialisation and unemployment on five men from the city.
- In the 1980s and 1990s the city had a reputation for gang crime linked to drugs, and particularly linked to firearms.

Large areas of derelict land, run-down housing and high levels of poverty added to Liverpool's poor image. Today, this reputation is less well deserved. Major regeneration of Liverpool's inner city has transformed the physical fabric of the city. Liverpool was European Capital of Culture in 2008. In the 2013 'Peace Index' — a crime survey — Liverpool was outside the top 20 least peaceful areas in England. On the other hand, in the government's 2016 'Happiness Survey' residents of Liverpool were the second least happy in the country.

Cities are complex places, and one part of the city may appeal to some people but not to others, as outlined in Table 30. Broadly speaking the inner city appeals to people at the early stages of their **life cycle**, but this declines with age. People from ethnic minorities may feel isolated in the white, middle-class-dominated suburbs. However, Oadby in Leicester is an example of an affluent suburb whose population in 2013 was 33% British Indian.

**Table 30** Perceptions of inner-city and suburban areas

| | People who like it: |
|---|---|
| **Inner city** | Recently arrived migrants: job opportunities are close by in the CBD and inner city housing is cheap; there may already be established ethnic communities.<br>Students: they are close to university, entertainment and most lack cars.<br>Young, professional workers: they can live in apartments close to work; all the entertainment facilities of the CBD. |
| **Suburbs** | Young families: the best schools tend to be suburban, houses have gardens and out-of-town retail parks are close by; ring roads and suburban rail networks make commuting to work relatively easy.<br>Older people and retired people: crime is usually low, it is more peaceful than the inner city; they generally don't use services in the CBD very much. |

> **Exam tip**
>
> As far as possible, try to back up your discussion of a place's image and perception of that place with data.

> **Exam tip**
>
> It is important to remember that, as with Liverpool, people's perceptions of a place can be very different to the reality of a place.

> **Life cycle** refers to the stages of people's lives: child, student, young single worker, raising a family, retirement. Places seen as desirable to live and work in change along this life cycle.

> **Knowledge check 42**
>
> In which area of the city are older people and retired people more likely to live?

# Perceptions of rural places

Rural areas are often perceived as 'ideal' places to live. Sometimes this is referred to as the 'rural idyll'. Many urban people have this attitude, which is based on:

- rural areas having picturesque landscapes of rolling hills and woodlands
- old, cottage-style housing with flower gardens
- a relaxed pace of life, free from stress and worries
- a strong sense of community, and activities like village cricket and socialising in a village pub
- places that are free of crime
- places that have a long history, and historic buildings like castles and village churches.

Media portrayals of rural places can reinforce this. TV series like *Emmerdale*, *Midsomer Murders* and *Heartbeat* are set in attractive countryside (even if the 'goings on' are sometimes murderous!). Rural places often brand themselves using art, television or literature to attract visitors. Examples include Hardy Country in Dorset, and recently Cornwall has begun advertising the locations in the series *Poldark* as tourist destinations.

However, as well as the rural idyll there is also a **rural paradox**. The reality of rural living may be more like that outlined in Table 31, especially in remote rural areas like Anglesey.

**Table 31** Life in rural areas

| Energy | Services | Housing |
|---|---|---|
| Many homes are not connected to gas pipelines, so have very expensive oil-fired boilers and central heating. | Post offices, shops, petrol stations and banks are often very limited — and more expensive than in rural areas. Schools and hospitals can be many miles away. | Houses are often old, with high maintenance costs and high heating costs. In National Parks, conservation rules can restrict improvements like double glazing. |
| **Transport** | **Population** | **Tourism** |
| Infrequent, expensive buses, high petrol/diesel costs and long distances to services all increase transport costs. | Ageing populations mean limited social opportunities for children and young people and a feeling of isolation. | Popular places can be swamped by summer tourists, but deserted in winter with seasonal services closed for months. |

It is important to recognise that just as cities have different areas within them (e.g. inner city, suburbs) so do rural areas. These areas are perceived differently:

- Remote rural: places to visit, but a very small number of people move there to retire, or 'get away from it all'.
- Accessible rural: popular retirement location, balancing the desire for rural peace and tranquillity with access to services in nearby market towns and cities; coastal places are very popular with retirees.
- Commuter villages: within a one-hour drive to a major city, these are popular locations for young families who are **counter-urbanising** and who commute to work.

The **rural idyll** is the perception of rural areas as peaceful, beautiful, relaxed and happy — the ideal place to live. It is just a perception; the reality may be very different.

**Exam tip**

You can use YouTube and other video sites to watch past episodes of television series and get a sense of how rural and urban areas are portrayed.

The **rural paradox** is the idea that some of the most desirable places to live in the countryside are also some of the least-well-served places in terms of services like healthcare, transport and shops.

**Knowledge check 43**

Name two household costs that are likely to be higher in rural areas compared to urban areas.

**Knowledge check 44**

Many people's perceptions of rural areas are based on a particular type of media representation. Which type?

**Counter-urbanisation** means moving out of the city or suburbs to live in a rural area.

## Evaluating diverse places

It is important to use data to determine whether people have a positive or negative image of a place. Much of the **quantitative data** that can be used are supportive of conclusions you might draw, as direct statistical evidence on image and perceptions are rare. Data that could be used include:

■ Census data about population growth and decline, age categories, ethnicity and health.

■ IMD data which specifically identify, at a small spatial scale, areas that are deprived and break this down further into seven deprivation domains.

■ Labour force surveys which tell us what average incomes in an area are, the types of jobs people do and whether they work full or part time.

These quantitative data are very useful as they provide a measure of the extent of social, economic and environmental problems in an area and can be used to compare places.

Other media, such as images, television documentaries, blogs and even art works can give contrasting evidence of a sense of a place. Some of these are controversial and need to be used with care.

# Why are there demographic and cultural tensions in diverse places?

■ Internal migration and international immigration have increased cultural diversity in the UK, even in some rural places.

■ Cultural diversity is reflected both in segregated places and places with distinctive characteristics that reflect the people living there.

■ Diversity can lead to tension and conflict between groups especially if change is rapid.

## A diverse UK

Since deindustrialisation began in the 1960s there has been significant internal migration in the UK which can be summed up as a 'North–South drift'. Internal migrants tend to be:

■ young, mostly under 35

■ relatively skilled/educated and motivated

■ seeking employment in an area of the UK that is more prosperous than the one they came from.

This means some regions gain, but origin regions lose their youngest and most talented people. Since 2002 about 2.7 million people moved from one local authority to another. Regionally the biggest 'losers' are the North East, West Midlands and Yorkshire and the Humber.

All regions of the UK have grown in population since 2004, but internal migration contributes to very different growth rates:

■ The East and South East grew by 0.8% each year between 2004 and 2014, and London grew by 1.8% annually.

■ Scotland, Wales, the North East and North West all grew by 0.5% per year or less.

---

**Exam tip**

Remember that living in a rural area can be very expensive, especially for lower income groups like young people and pensioners.

**Quantitative data** are numerical data which have set values, i.e. measures the quantity of something.

**Knowledge check 45**

What is meant by counter-urbanisation?

**North–South drift** refers to the movement of people within the UK from places in the North towards the London, the South and South East.

Because most internal migrants are young, this has contributed to some quite significant differences in average age between regions:

- London is youngest, with an average age of 34.
- In Scotland, the North East and North West the average age is 40–41.

The South West is an anomaly as it has the highest average age at 42.9 but also has positive **net internal migration**. This is because it is a very popular retirement destination.

Internal migration is not the only component of population change; there are three. Figure 28 illustrates this. Changes as a result of international migration have been very significant in the last 50 years, but have impacted on some places more than others. Immigrants to the UK have tended to live in some places more than others, especially some cities.

Growth or decline in population, as well as average age and ethnicity, depends on the balance of:

**Figure 28** Components of population change

Immigrants who arrived in the 1950s, 1960s and 1970s are often called **post-colonial migrants**. Most have children and grandchildren born in the UK. These second- and third-generation populations tend to live in similar places to where their parents and grandparents first settled (Table 32). The largest immigrant groups in the UK are from the Indian sub-continent (India, Pakistan and Bangladesh).

**Net migration** is the balance of people arriving in and leaving an area in a year. More arrivals than leavers mean positive net migration; more leavers than arrivals mean negative net migration.

**Knowledge check 46**

Which UK region has the youngest average age?

**Post-colonial migrants** arrived from countries that were once colonies of the UK but became independent after 1947 (sometimes called the 'New Commonwealth').

**Table 32** Examples of UK immigrant groups

| Who | When and why | Major concentrations |
| --- | --- | --- |
| Afro-Caribbean, especially from Jamaica and Trinidad & Tobago | Beginning in 1948 and peaking in the late 1950s<br>Filled a post-WWII labour shortage in the UK | London (Brixton, Lewisham) and Birmingham (Handsworth, Aston) |
| Indian | 1950s–1970s<br>Economic migrants seeking work in the UK, many in factories | London (Harrow, Hounslow, Brent), Wolverhampton, Coventry, Leicester, Blackburn |
| Pakistani | Mainly in the 1950s and 1960s; many were well-qualified and skilled from cities in Pakistan | London (Ilford, Barking, Walthamstow), Birmingham, Bradford and Manchester |
| Bangladeshi | Mainly in the 1970s | London (Tower Hamlets and Newham) and Birmingham |
| Somalis (East Africa) | 1988–2009<br>About 140,000 Somali-born people are resident in the UK, most arriving as refugees fleeing war | London (Southall) |
| EU A8 | 2004 onwards<br>Economic migrants across a wide range of skill levels | Widely distributed across the UK, much more so than the other types of immigration above |

There are exceptions to the general pattern that immigrants settle in urban areas. Since 2004 some **A8 migrants** from the EU have settled in more rural locations because there were particular skills shortages they could easily fill:

- Northern Scotland, working in the fishing, fish processing and fish packaging industries.
- Lincolnshire and Cambridgeshire, working in farming, food processing and packaging.

Boston in Lincolnshire had 16% of its resident population from A8 countries in 2015. This is a very high percentage, up from basically 0% in 2004, and represents a very rapid change to the demography and culture of a small market town. Large-scale migration will create a number of social challenges in rural areas:

- Housing shortages and price rises, because the amount of housing available is limited.
- Challenges of delivering education and healthcare, because of language barriers and limited service supply.
- Cultural challenges in traditional rural areas that rarely experience 'outsiders' or change.

However, A8 migrants also reduce the average age of rural areas and boost population because they tend to be young and have children. Economic opportunities are created because many A8 migrants have set up their own businesses.

## Segregation and variation

Many immigrant groups tend to live in clusters, i.e. they are segregated as we saw in Table 27. This segregation has social and cultural dimensions as well as an economic dimension. In 2014 there were estimated to be up to 150,000 Russians living in London. Most live in the wealthiest locations such as Kensington and Chelsea because they have 'exported' wealth from Russia to invest in UK property.

Conversely, the British Bangladeshi population is concentrated in some of the poorest parts of the UK:

- Around 33% of the population of the London borough of Tower Hamlets (ranked 247 worst out of 326 for deprivation in England) is Bangladeshi, and this rises to over 50% in the wards of Whitechapel and Spitalfields.
- In 2015 about 50% of British Bangladeshis were born in Bangladesh, much higher than for British Indians and Pakistanis, and 50% of British Bangladeshis speak Bengali as their first language.
- Over 65% of Bangladeshis live in low-income households and the average household size of five is much higher than the UK average.
- Bangladeshis have the highest levels of illness of any UK immigrant group.
- The unemployment rate for British Bangladeshis aged 16–24 was 46% in 2014 compared to 19% for white British young people.

Lack of skills, inability to speak English and discrimination in the jobs market explain why British Bangladeshis tend to live in deprived areas. On top of this is the tendency for them to experience segregation on cultural grounds. However, it is not all negative news for British Bangladeshis. By 2013 the percentage of 16-year-old Bangladeshis achieving five or more grade Cs at GCSE had surpassed both the British Pakistani and white British level.

**Exam tip**

Learn when different groups of immigrants arrived in the UK so you have an accurate timeline.

**A8 migrants** are those from eight eastern European countries that joined the EU in 2004 (Poland, Latvia, Lithuania, Estonia, Hungary, Czech Republic, Slovakia and Slovenia).

**Knowledge check 47**

When did post-colonial migration to the UK begin?

**Exam tip**

Make sure you have some data to back up any explanation of segregation.

Given the concentration of ethnic groups in some locations, it is no surprise that the urban landscape of some towns and cities has been altered to reflect the social characteristics and culture of dominant ethnic groups. This includes:

- places of worship such as mosques and Hindu temples
- shops selling cooking ingredients for specific cultures, and traditional dress
- community centres and sports and leisure facilities to suit different ethnic groups.

Many places in the UK have an urban landscape that looks very different to what one might expect because of the large number of ethnically specific retail outlets. Examples include Brick Lane — also known as 'Banglatown' — in London's Tower Hamlets, which is famous for its curry restaurants.

It is fair to say that the first post-colonial migrants who arrived in the UK in the 1940s and 1950s experienced outright hostility and widespread discrimination from a section of the UK's population. This has not gone away, but the experiences of ethnic groups has changed over the decades and their perceptions of where they live has also changed. This is due to a number of trends:

1 Ethnic communities have grown wealthier over time, by setting up businesses and moving into professional and managerial jobs.

2 Second- and third-generation immigrants have gone through the UK education system and their culture is more likely to be a hybrid of British and Asian, or British and Caribbean.

3 Communities have put their cultural 'stamp' on the built environment, giving the areas they live in a cultural familiarity they did not have decades ago.

4 Members of immigrant communities have become local councillors and MPs, giving their people a voice they lacked decades ago.

Table 33 outlines some examples of the cultural evolution that has taken place in the UK since the 1950s.

**Table 33** Cultural evolution in the UK

| | |
|---|---|
| **Suburban communities** As the economic prosperity of Britain's Indian community has grown they have moved out of poor inner-city areas and into the prosperous suburbs, such as Oadby in Leicester which was 33% Indian by 2011. | **Cultural festivals** The Notting Hill Carnival (begun in 1966) has been joined by numerous Hindu melas (meaning a fair, or gathering) in UK cities as celebrations of ethnic culture welcoming everyone. |
| **Political representation** In 2015, 41 MPs from ethnic minorities were elected, up from 4 in 1987. Ethnic minorities make up 4% of local councillors. In both cases this is far short of the 13% of the UK population who were from an ethnic minority in 2015. | **Cultural hybridisation** Some aspects of ethnic minority culture such as curries, bhangra and reggae music have become part of British culture — and in the process have been altered by being in Britain. |

**Knowledge check 48**

Which ethnic minority group is concentrated around Brick Lane in Tower Hamlets?

**Cultural hybridisation** happens when aspects of one culture are affected by another's, producing a new variant. It is very common in food, music and fashion.

**Exam tip**

Exam questions will want you to discuss the positive impacts of the UK's increased cultural diversity as well as the problems, in a balanced way.

# Tension and conflict

As we have already seen, Southwark in London has a culturally diverse population. It also suffers from high levels of deprivation. However, property is expensive and in high demand. Southwark is very close to central London and for many transnational corporations (TNCs) it is an ideal place to either locate or invest. Southwark is subject to a number of ongoing regeneration plans:

- Regeneration of the Aylesbury Estate, a 1960s tower block council housing estate with a reputation for poverty and crime; regeneration began in 2009.
- A similar regeneration of the Heygate Estate; this was demolished between 2011 and 2014.
- Plans to demolish and regenerate the Elephant & Castle shopping centre.

East Walworth, the ward containing the Heygate Estate, was 48% white, 31% black and 10% Asian in 2011. As a deprived area, many of the people live in council housing and are on very low incomes. Local opposition to the plans has been organised by groups such as Just Space and 35percent. Their opposition is focused on:

- Lack of community involvement in the regeneration plans, and forced relocation.
- Southwark Council selling off land that was once council housing to private developers like Lend Lease.
- The lack of affordable housing, which is likely to be 25% of the regenerated housing not the 35% which is Southwark Council's target.
- The risk that this part of London will be **gentrified**, or even **'socially cleansed'**.

Those in favour of the schemes argue that they will create jobs, improve the urban environment and improve the reputation of the areas. This example shows that the priorities of local and national government (regeneration), TNCs (profit) and local community groups do not always align.

In 1950 the UK was almost 100% white British, but today it is around 81% white British. This very large increase in cultural diversity has happened relatively smoothly, but not entirely so. Although rare, there has been a history of racially motivated riots in the UK, including:

- the Notting Hill (London) riots in 1958
- the Brixton riots (London) in 1981
- the Broadwater Farm riot (London) in 1985
- the Bradford riots in 1995 and 2001.

Riots are complex and almost always occur in deprived areas. This means issues like poverty, deprivation and lack of opportunity are often the root cause of people rioting. However, in some cases rapid population and cultural change seem to contribute. Often riots are 'triggered' by a particular incident — such as an arrest or assault — that is interpreted very differently by opposing communities.

There are also examples of so-called '**white flight**' and in some cases accusations that an existing population has been 'forced' out of an area by an incoming ethnic group. This is the case in the Bury Park area of Luton (Figure 29), once a white working-class area that is now predominantly Muslim.

Areas are **gentrified** (gentrification) when wealthy people move into a low income, working-class area, which leads to higher house prices.

**Social cleansing** implies a particular section of a society being removed from an area; in a UK context it really means being priced out of an area because of rising housing costs.

### Knowledge check 49

What is the name for the process when a low income area receives an influx of new, wealthy residents?

**White flight** is when an existing white population leaves an area because an ethnic minority group begins to move into it.

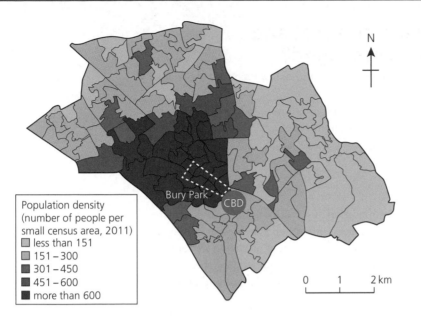

**Figure 29** Population density of Asian British people in Luton 2011
Source: www.luton.gov.uk and ONS

Luton illustrates why cultural change can sometimes lead to hostility between long-term residents and recently arrived immigrants. Its built environment has changed:

■ The Asian ethnic group (the majority of whom are British Pakistani or Bangladeshi) are unusually highly concentrated in one area of Luton (Figure 29).
■ Bury Park has over 20 mosques, for many different Muslim sects.
■ There are also madrassas — Muslim schools — often but not always dedicated to religious study.

Mosques and madrassas are not well understood by non-Muslims. In some cases they have been linked to Muslim extremism (a charge many Muslims consider unfair). Luton was identified in a 2016 BBC report as the second largest source of jihadists, after London, travelling to Syria in support of ISIS. Luton has also seen demonstrations and clashes involving far-right groups such as the English Defence League.

It is perhaps fair to say there are extreme positions on both sides in Luton. A minority of white British people feel the scale and pace of ethnic change in Luton has threatened their culture, and a minority of Muslims take an extreme religious position and support causes that the majority of UK residents do not. It should also be said that Figure 29 hints at **social exclusion**. Luton's Asian population is significantly more deprived than either the local black or white population. It may be the case that Luton's Asian population feel excluded and marginalised economically as well as culturally. That is not a recipe for good relations in a culturally diverse population.

**Exam tip**

Remember that the majority of people do not hold extreme views, and are not involved in conflicts and tension between different community groups.

**Social exclusion** is when people, or whole communities, feel marginalised and blocked from the opportunities wider society enjoys.

**Knowledge check 50**

Luton's Asian British population is concentrated in one area. What name can be given to this?

# How successfully are cultural and demographic issues managed?

■ The problems resulting from rapid demographic and cultural change are managed in different ways.

■ The success of management in urban and rural areas, to reduce inequality and social problems, can be measured using different criteria.

■ Different stakeholders have different perceptions of issues, and may use different criteria to judge their success.

## Measuring progress

In areas that have experienced rapid immigration and cultural change, one way to measure how successfully immigrants have been integrated is to look at employment levels. Southwark appears to do well in terms of employment as shown in Table 34. Although unemployment is a little higher than the rest of London and the UK, earnings, economically active (as a % of the working age population) and professional and managerial jobs are all higher.

**Table 34** Income and employment data 2016

|  | Southwark | London | UK average |
|---|---|---|---|
| Economically active (%) | 83 | 78 | 78 |
| Unemployed (%) | 7 | 6 | 5 |
| Professional and managerial jobs (%) | 41 | 36 | 30 |
| Weekly earnings, full-time workers (£) | 686 | 660 | 530 |

However, the data in Table 34 are an average for the whole of Southwark. The proportion of the black British population that is economically active is about 65%, much less than the average. Across London, the proportion of people earning less than the **London Living Wage** varies by ethnic group (Figure 30). It is particularly poor among the Pakistani and Bangladeshi ethnic groups, and has worsened since 2007–08.

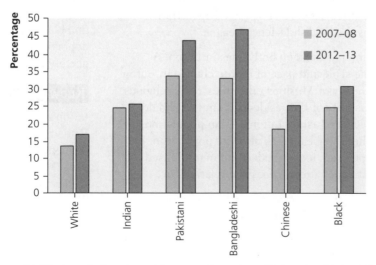

**Figure 30** Workers in London with wages less than the London Living Wage
Source: www.trustforlondon.org.uk

The **London Living Wage** is the hourly rate of pay estimated to provide a decent income for people in London, i.e. they could afford to live reasonably comfortably. In 2015 it was £9.40 per hour.

**Knowledge check 51**

On Figure 30, which ethnic group saw the largest deterioration in their pay between 2007–08 and 2012–13?

The proportion of 16-year-olds who did not achieve five GCSEs (at any grade) in Southwark fell from 18% in 2000 to 13% in 2007, but this is much higher than the 9% UK average. Around 42% of Southwark's 19-year-olds have no qualifications, compared with only 36% for London as a whole. Southwark has one of the highest rates of infant deaths before age one, and a high rate of homelessness. In Southwark:

- Young, educated migrants from other parts of the UK, i.e. internal migrants, appear to be doing very well.
- Some international migrants from the EU and skilled professionals from other countries are prospering.
- Some ethnic minority groups in Southwark suffer from poverty and unemployment.

## Social progress and cultural assimilation

There is some evidence of **social progress** in Southwark, but also very stark inequalities. Almost 86% of people in Village **ward** (see Figure 26) are in the least deprived category nationally, whereas 75% of people in Livesey ward are in the most deprived category nationally — despite the two wards being only 3 km apart. Village ward had a male life expectancy of 84 years (20% from ethnic minority groups) in 2013 compared to 79 in Livesey (62% from ethnic minority groups) and 74 in Camberwell Green (61% from ethnic minority groups). Life expectancy and ethnicity appear to be linked.

It is interesting to look at how the Index of Multiple Deprivation (IMD) changed between 2010 and 2015 in Southwark's Village and Livesey wards:

- In Village, the IMD score improved from 14.9 to 12.3, a fall in deprivation of 2.6.
- In Livesey, it improved from 39.3 to 38.9, a fall of 0.4.
- Two **small census areas** in Livesey actually had worse deprivation in 2015 than 2011, whereas all areas in Village improved.

Deprivation fell in Southwark 2010–15, but more in the least deprived wards. The gap between the least and most deprived in Southwark actually increased.

Immigrant communities undergo a process of **cultural assimilation** over time, especially second and third generations. Experience of UK language, media, education and employment should help immigrants 'fit in' over time. This is not to say traditions and culture are necessarily lost, but they are altered. There is some evidence that ethnic groups do become more involved in society over time. The election for the Mayor of London in 2012 saw a wide variation in voter turnout in Southwark:

- 52% in Village, a mostly white British ward
- 37.6% in Lane ward, which has an established black British Caribbean community
- 33.8% in Livesey ward, with its large black British African community
- 30% in Cathedrals, a ward with a high percentage of recently arrived Asian ethnic groups.

Southwark hosts London's Africa Centre. This NGO and community group was originally founded in 1964 and moved its HQ to Southwark in 2013. There are other specific groups such as the Organisation of Blind Africans and Caribbeans founded in 1988 and the Ethnic Health Foundation set up in 2002 to support health among black and other ethnic minority groups.

**Exam tip**

It is very important to recognise that in an inner city area like Southwark, some people and areas are very prosperous but others are not.

**Social progress** means how a community improves its quality of life, health and welfare over time.

A **ward** is a Census area sub-division; most have a population of around 10,000–15,000.

**Small census areas** (technically lower super-output areas) are small area divisions of the Census, usually containing about 1500 people.

**Knowledge check 52**

Which ward improved its deprivation level the most between 2010 and 2015, Village or Livesey?

**Cultural assimilation** is the process whereby the culture of one group gradually begins to resemble that of another group. New immigrant groups slowly become more similar to the society they have moved into. It is a two-way process, e.g. UK people adopting the foods of immigrant groups.

If ethnic minority groups and the wider society of the UK are living more peacefully together, we might expect to see a fall in **hate crimes** over time. Table 35 shows hate crime recorded by the Metropolitan Police in London 2008–15. It appears to show a fall up to 2011–12, but a sharp increase in more recent years. This is worrying: however, changes in crime statistics methodology and greater reporting of hate crime (i.e. better awareness of it) could affect these statistics.

**Hate crime** is crime motivated wholly or in part by the victim's membership (or assumed membership) of a racial group or a religious group.

**Table 35** Hate crime in London 2008—15

|  | 2008–09 | 2009–10 | 2010–11 | 2011–12 | 2012–13 | 2013–14 | 2014–15 |
|---|---|---|---|---|---|---|---|
| **Racist incidents** | 10,190 | 10,541 | 9,405 | 7,983 | 9,453 | 9,749 | 11,540 |
| **Religious hatred** | NB Hate crime was recorded differently in 2012–13 | | | 607 | 634 | 915 | 1371 |

## Urban stakeholders

In culturally and demographically dynamic communities such as Southwark, change is ongoing. Change tends to be contested, because the priorities of local communities and national and local councils do not always align. Table 36 summarises some initiatives in Southwark, by different **stakeholders**, and assesses their impact.

> **Exam tip**
>
> Crime statistics are difficult to interpret and use. Do not be afraid to comment on the fact that some data are easier to use and more reliable than other data

> **Exam tip**
>
> Try to be specific in the exam when you are discussing different ethnic groups. Terms like 'Asian' are often too generalised.

**Stakeholders** are any groups or individuals involved in, or interested in, regeneration. They range from residents, environmentalists and businesses to local councils and planners.

**Table 36** Stakeholder programmes in Southwark

| Stakeholder | The initiative | Assessment |
|---|---|---|
| PwC is a major global accountancy and auditing TNC which has a large office in the 'MORE London' regeneration area of Southwark. | Between 1998 and 2006 PWC invested £1.8 million and 30,000 employee volunteer hours supporting community education in Southwark's schools including mentoring, skills workshops and other support, plus supporting out-of-work young people in partnership with the Prince's Trust NGO. | The support of local businesses is important, especially for young people, in terms of raising aspirations and developing skills. |
| The UK Government is responsible for national counter-terrorism and anti-extremism policies. | The government's counter-terrorism strategy is called 'Contest', and 'Prevent' is the part of it designed to stop young Muslims being radicalised. In 2015 Southwark was added to the list of areas receiving special support. Teachers, community workers and even places of worship have a duty to work to prevent radicalism. | Prevent has been controversial, because of its focus on one religion — Islam. Critics argue it risks alienating the very groups most at risk from Muslim extremism. |
| Southwark Council is one of the main providers of housing in the area. | Regeneration plans for council housing estates including the Aylesbury Estate and Heygate Estate have involved selling off council housing and regenerating the areas as joint ventures between housing associations (affordable housing) and private property developers. | Many local residents' groups have protested strongly against Southwark Council's plans, arguing housing association rents will be higher and people are being forced out of their homes. |

Different stakeholders will use very different **criteria** to judge the success of the programmes in Table 36. The regeneration of the Aylesbury and Heygate council estates could be judged using criteria such as:

- The number of affordable homes completed.
- The profits made by the private developers when selling new houses.
- Improvements to the built environment such as the number of trees, or proportion of open space in an area.
- Rising property prices.
- Trends in deprivation, using the IMD, or trends in crime rates.
- Demographic trends such as the number of single person households versus families.

It's likely that existing residents of the Aylesbury and Heygate council estates will be more interested in rent levels and housing quality rather than the profits of the private developers. Young professionals moving to Southwark may benefit as more apartment-style housing becomes available to them because regenerated former council estates contain a mix of property to buy, as well as affordable homes to rent.

Some people in Southwark may feel that their **lived experience** of a place is at risk due to change. An example is the Canada Estate council housing in Rotherhithe. Southwark Council plans to regenerate these homes but there is no guarantee that existing residents will be allowed to stay. The quotes below are from two existing residents of the Edmonton flats:

- 'I'm now 90 years old. I came over on the *Windrush*, served my country and have suffered from cancer, but I still clean my windows and look after my home. I want to be buried on this estate.' Eric Paule-Drysdale, 90, a British black Caribbean man who emigrated to London onboard the *Empire Windrush* ship in 1948.
- 'I have never seen a community like this, everyone is friendly and we feel very safe. I am very sad if I have to move somewhere else. I am part of this community and I am happy here.' Jixu Ma, originally from China.

Both of these quotes are from different ethnic minority groups; people who have made their home in Southwark but feel that it is now at risk.

## Rural stakeholders

Rural places in the UK rarely have culturally diverse populations; as we have seen there are a few exceptions where A8 migrants have moved to rural Lincolnshire and the Highlands of Scotland. Rural places do have demographically diverse populations with young people, working people and retired people.

Anglesey, in 2004, was described in this way in a report on its economic development:

Anglesey is an island with deep-seated economic problems: high levels of migration of its young people, high levels of unemployment, significant levels of social deprivation, rural economic problems and a number of small-scale, low skill employment sectors.
Source: anglesey.gov.uk

**Knowledge check 53**

Name a major TNC that has invested in community initiatives in Southwark.

**Exam tip**

You do need examples of specific strategies and schemes that have changed, or are changing, both rural and urban places.

**Criteria** are the standards or measures people use to judge whether something is a success, or not.

**Lived experience** refers to people's feelings about a place, what is important to them and what a place means to them. This is strongly related to a place's traditions, its community spirit and its history.

- This type of statement is likely to resonate with Anglesey's young adults, who tend to leave the island to look for work.
- Immigrants, both internal and international, are unlikely to see the statement as a reason to move there.
- Older people, and those looking to retire, might be less concerned and focus more on the island's landscape and tranquillity than its economic issues.

Anglesey has been affected by a number of local and national changes in the last few decades:

- The increase in leisure time and personal mobility through mass car ownership has boosted tourism on Anglesey, bringing £250 million to the island economy each year.
- However, tourism is highly seasonal with most visitors arriving from May to September, meaning jobs are also seasonal.
- Visitors have high expectations: *a five-star island with three-star facilities*' was one recent description of Anglesey, hardly a ringing endorsement.
- Anglesey's isolation means it is not popular with day-trippers, but it is popular as a **second home** destination: there are about 2300 second homes on the island. In 2016 nearly 800 of these had not been occupied for more than a year.
- Second home demand pushes up house prices, meaning more reasons for the young to leave.
- Anglesey's nuclear power station, Wylfa, opened in 1971 but closed in 2012 with no plans for a replacement (the UK has not invested in new nuclear power since the early 1990s).

Anglesey's future may look relatively bleak, but there are some plans to revitalise the island outlined in Table 37.

**Table 37** Major developments on Anglesey

| Energy Island | Anglesey Council's plan is to create an economic hub on the island focused on low-carbon energy. There are plans for biomass, tidal and new nuclear developments as well as a Science Park focused on energy research. | Plans for a new nuclear power plant were put on hold in 2012, dealing a blow to Energy Island. The anti-nuclear power group 'People Against Wylfa B' led opposition to a replacement nuclear plant including a protest by 300 people in 2012. |
|---|---|---|
| Council tax on second homes | The Welsh Government voted to allow local councils to increase council tax by 100% on second homes from 2012 onwards, and Anglesey Council may start with a 25% increase. | This policy aims to reduce second homes and empty homes, so increasing housing supply for local young people, but they will also need decent jobs. |
| Land & Lakes development | A £150 million private leisure park development at Penrhos near Holyhead, with 300 homes and 300 holiday lodges, is to be built on the former site of the Anglesey Aluminium works. | The scale of the plans has upset some locals, and £20 million will have to be paid to Anglesey Council to offset the costs of new infrastructure and services for the development. It does not diversify the local economy, but will provide some jobs. |

**Second homes** are owned by non-residents of an area, and used for only part of the time, i.e. holidays or the summer only.

**Exam tip**

Remember that Anglesey, like most rural areas in the UK, is not ethnically diverse but it does have diversity in terms of population age groups.

**Knowledge check 54**

Which industry is worth £250 million each year to Anglesey?

Anglesey's problems are not unusual in UK rural areas. In 2016 Anglesey Council faced a £5.6 million budget shortfall, and was planning to increase council taxes by 4.5%. Higher local taxes and fewer local services will not attract people or businesses. Many people are attracted by the image of the place (beautiful coastline and countryside), but the reality (lack of jobs and housing, isolation) is very different. Young people and older people want very different things from Anglesey (Table 38).

**Knowledge check 55**

Which group of people are affected the most by the lack of affordable housing on Anglesey?

**Table 38** Different wants and needs of young and older people from Anglesey

| Younger people | Older people |
|---|---|
| ■ Jobs that are well paid, not seasonal and have promotion opportunities.<br>■ Connections to mainland Wales and beyond, which are fast and affordable.<br>■ High quality education and skills training.<br>■ Affordable housing.<br>■ Social and leisure activities. | ■ An attractive landscape that is protected from further development.<br>■ Limited new housing development, which would take up land, increase population, and lower existing house prices.<br>■ Key services, especially health services. |

## Summary

- Population growth, density and population structure in terms of age categories, varies widely across UK regions with some areas growing faster and with younger populations.
- Some places, especially cities, are locations of cultural diversity and these are often changing because of internal and external migration.
- All places are changing as a result of local, regional, national and global forces — but change is more rapid in urban areas compared to rural areas.
- Cities, and areas within them, are perceived very differently by people of different ages, ethnicities and life cycle stages.
- Rural places often have an idyllic image, but in reality they can be challenging places to live especially for older people and people on low incomes.
- Internal and international migration has had a profound impact on the characteristics and culture of urban places such as Southwark.
- In many urban areas people are clustered according to income level but also ethnicity, which gives areas a particular character.
- Rapid cultural change can lead to tensions in an area, as new people move in and the population and physical environment of places changes.
- Even in a small area, such as Southwark, evidence of social progress reveals stark differences between locations and different groups.
- In both rural and urban places, attempts to manage cultural and demographic change are viewed very differently by different stakeholders who use different criteria to measure success.

# Questions & Answers

## ■ Assessment overview

In this section of the book, two sets of questions on each of the content areas are given, one set for AS and one for A-level. For each of these, the style of questions used in the examination papers has been replicated, with a mixture of short answer questions, data stimulus questions and extended writing questions. The relative proportions and weightings of the marks varies between AS and A-level.

All questions that carry a large number of marks (at AS and A-level) require candidates to consider connections between the subject matter and to demonstrate deeper understanding in order to access the highest marks. At **AS** the breakdown of the questions per topic is:

- Globalisation (28 marks)
  Typical question sequence: 1, 2, 3, 4, 6 and 12 mark questions
- **And either** Shaping places: Regenerating places (28 marks)
  Typical question sequence: 1, 2, 3, 4, 6 and 12 mark questions
- **Or** Shaping places: Diverse places
  Typical question sequence: 1, 2, 3, 4, 6 and 12 mark questions.

There are also fieldwork skills questions on Regenerating places and Diverse places at AS. The overall paper takes 1 hour and 45 minutes and is worth a total of 90 marks, making up 50% of the AS qualification. Fieldwork skills are covered in Student Guide 4 on Geographical skills. There are no fieldwork skills questions on the A-level exam paper.

At **A-level** the breakdown of the questions per topic is:

- Globalisation (16 marks)
  Typical question sequence: 4 and 12 mark questions
- **And either** Shaping places: Regenerating places (35 marks)
  Typical question sequence: 6, 6 and 20 mark questions
- **Or** Shaping places: Diverse places (35 marks)
  Typical question sequence: 6, 6 and 20 mark questions.

Overall the A-level paper takes 2 hours and 15 minutes and is worth a total of 105 marks, making up 30% of the A-level qualification.

In this section of the book, each of the content areas is structured as follows:

- sample questions in the style of the examination
- Levels-based mark schemes for extended questions (6 marks and over) in the style of the examination
- example student answers at the upper level of performance
- examiner commentary on each of the above.

Carefully study the descriptions given after each question to understand the requirements necessary to achieve a high mark. You should also read the commentary with the mark schemes to understand why credit has or has not been awarded. In all cases, actual marks are indicated.

# ■AS questions

## Globalisation

### Question 1

**(a)** Define the term 'outsourcing'. (1 mark)

**(b)** Study Figure 1.

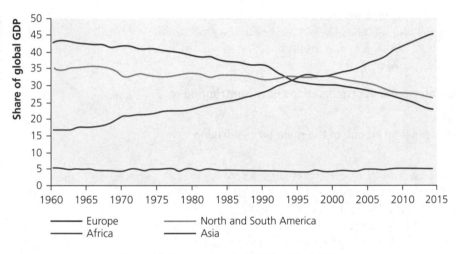

**Figure 1** Share of global GDP 1960–2015

**(i)** Calculate the change in Asia's share of global GDP between 1960 and 2015. (1 mark)

**(ii)** Name the global region with the largest decrease in global share of GDP between 1960 and 2015. (1 mark)

**(iii)** Suggest one reason for the rapid increase in Asia's share of global GDP. (3 marks)

ⓔ Part (a) is a simple recall question. Part (b)(i) requires use of Figure 1 to identify the relevant values from the y-axis for the two dates, followed by a simple subtraction to calculate the change. Part (b)(ii) requires the graph to be used again, taking care to differentiate between the declines of Europe and North America. Part (b)(iii) needs an explanation but because 3 marks need to be gained from one extended point, this needs to be in depth. An example could be used as part of this answer.

# Questions & Answers

**(a)** Companies transferring work to outside suppliers, often overseas.

**(b) (i)** 17% in 1960, 45% in 2015 = 28% change

**(ii)** Europe

**(iii)** Asia's impressive growth can be put down to the global shift of manufacturing industry from developed regions like Europe, to China and South Korea. This has created millions of new jobs, and increased exports and incomes, all contributing to a larger share of global GDP. This has been led by TNCs like Apple that have shifted production to China.

ⓔ **6/6 marks awarded** This is a good answer. Part (a) scores 1 mark for a correct definition. In Part (b)(i) and (b)(ii) Figure 1 has been used accurately so both score 1 mark each. The answer to (b)(i) shows some of the working for the answer, which is good practice. Part (b)(iii) takes the single idea of the global shift and extends using the example of Apple, and explanations of rising incomes and exports leading to greater GDP, so scores 3 marks.

**(c)** Explain two ways in which international organisations have contributed to globalisation. (4 marks)

**(d)** Explain why TNCs are often considered one of the main forces driving globalisation. (6 marks)

ⓔ Part (c) requires two reasons, each with an extension to gain 2 × 2 marks. Be careful not to write in detail about only one reason here. Part (d) is a more extended question. It is about TNCs so a detailed explanation is needed with reference to TNCs and good use of terminology to gain full marks. The phrase 'one of the main' does allow you to mention other factors, but this should not be the main focus of the answer. With questions for 6+ marks using examples to add depth is always a good idea. The Levels mark scheme for a 6 mark question is shown below. The same Levels mark scheme is used for the 6 mark questions in the AS options questions.

| Level 1<br>1–2 marks | ■ Demonstrates isolated elements of geographical knowledge and understanding, some of which may be inaccurate or irrelevant.<br>■ Understanding addresses a narrow range of geographical ideas, which lack detail. |
|---|---|
| Level 2<br>3–4 marks | ■ Demonstrates geographical knowledge and understanding, which is mostly relevant and may include some inaccuracies.<br>■ Understanding addresses a range of geographical ideas, which are not fully detailed and/or developed. |
| Level 3<br>5–6 marks | ■ Demonstrates accurate and relevant geographical knowledge and understanding throughout.<br>■ Understanding addresses a broad range of geographical ideas, which are detailed and fully developed. |

**Student answer**

**(c)** The World Trade Organization has, since the 1950s, held trade talks aimed at reducing import and export tariffs and taxes, which has helped increase globalised trade as volume rises as taxes and tariffs fall. The International Monetary Fund has encouraged developing countries to accept foreign direct investment, which in turn has allowed TNCs to expand into new markets in Africa and Asia.

**(d)** TNCs are often global companies with operations world-wide. An example is Apple. It is based in California where research and development on new products is done, but products are made by third-party suppliers such as Foxconn in China, and sold in Apple stores world-wide. This global network of production connects places through trade and even the movement of workers. In addition, ICT companies like Apple and Google actually create new technologies that connect people and contribute to time-space compression such as internet apps, mobile phones and social networking. Other TNCs such as Boeing built the jet aircraft that reduce the friction of distance. Global TNC brands are known world-wide and contribute to cultural globalisation and global media. TNCs such as Disney and CNN spread western media to a global audience. However, TNCs operate within a global economic system which has been influenced by greater free-trade through the work of the World Trade Organisation, and TNC foreign direct investment is encouraged by national governments — such as China's special economic zones. TNCs are crucial to globalisation but they are not the only drivers of it.

@ **10/10 marks awarded** The answer to Part (c) scores 4 marks because it explains the role of two different international organisations, and in each case provides an extended explanation of how each action has contributed to globalisation, i.e. increased global trade and expansion of TNCs. Four marks could also be gained by a very detailed answer of how one organisation has contributed in two ways.

The answer to Part (d) scores 6 marks. It uses examples of TNCs to develop ideas, and covers more than one aspect of TNCs, i.e. their role as global networks as well as how they develop technology and spread cultural ideas. The answer shows deeper understanding by recognising that TNCs are not the only drivers of globalisation, but they are a very important element of globalisation. These 6 mark questions do provide the opportunity for a slightly extended answer that can show the breadth of your understanding as well as depth.

## Shaping places: Regenerating places

## Question 2

(a) Define the term 'farm diversification'. (1 mark)

(b) Study Figure 2.

(i) Calculate the difference in visitor numbers to the Eden Project between 2002 and 2015. Show your working. (2 marks)

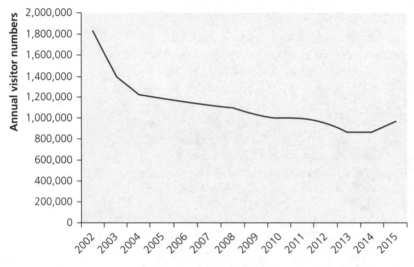

**Figure 2** Annual visitor numbers to the Eden Project, Cornwall, a major rural regeneration project and tourist attraction that opened in 2001

(ii) Suggest one impact on the rural area of the change in visitor numbers shown on Figure 2. (3 marks)

ⓔ Parts (a) and (b) are a combination of recall (a) and interpretation and explanation of a data stimulus resource (b). Figure 2 needs to be studied carefully taking note of the overall decline, but also the slight increase in visitor numbers between 2014 and 2015. Part (b)(ii) requires an extended explanation of one impact. This is likely to focus on negative economic impacts because of the large decline in visitor numbers, although a positive impact such as less congestion on rural roads could also be used.

---

**Student answer**

(a) Farms generating income from new business activities.

(b) (i) 1.83 million in 2002 minus 0.96 million in 2015 = 0.87 million difference

(ii) The number of visitors to the Eden Project has roughly halved since 2002, which could mean fewer people are employed there, so local unemployment may be higher than it was in 2002 so there is less money in the local economy from employees and visitors.

---

**ⓔ 6/6 marks awarded** The answer to Part (a) is correct. Any answer indicating an income source other than farming would gain 1 mark. The student correctly identifies the values for the two years from Figure 2 for Part (b)(i) and shows these in their answer, as well as working out the difference between them correctly, scoring 2 marks. Mark scheme questions such as this have an 'acceptable' range of correct answers usually +/– 5% around the precise answer. Notice that in Part (b)(ii) the basic impact — loss of jobs — is followed by two further points related to this reason, so this answer scored 3 marks.

**(c)** Explain two criteria that can be used to judge the success of regeneration schemes. (4 marks)

**(d)** Explain the importance of re-imaging as part of the wider regeneration process. (6 marks)

ⓔ Part (c) is a point-marked question. Care needs to be taken to choose two different criteria, such as deprivation levels and demographic change. In addition 2×2 developed explanations are needed rather than a long explanation of only one criterion. Part (d) is a Level-marked question (see Levels mark scheme on page 76). Answers need to demonstrate an understanding of re-imaging, but also how this fits into a broader physical and economic regeneration strategy. An example could be used to illustrate this answer.

### Student answer

**(c)** The Index of Multiple Deprivation could be used to judge the success of regeneration schemes because changes in the income or crime domains between two dates, such as 2010 and 2015, could show improving economic or social conditions as a result of regeneration. Population numbers from the census could be used as a growing population indicates a successful area attracting people and investment, especially if growth is among younger working age people aged 20–30.

**(d)** Regeneration involves not only rebuilding infrastructure and developing new residential and commercial buildings but also changing the image of a place. This re-imaging aims to change the external perceptions of a place, which changes people's view of the area and helps attract investment. Both in Liverpool and Salford Quays, regeneration has involved re-imaging to associate the areas with arts and culture, such as the Lowry and MediaCity in Salford Quays, to change the perception away from an image of a run-down deindustrialised area. In some cases the use of branding and logos is important for use on advertising. However, on its own a new image does not lead to successful regeneration because it cannot improve quality of life for existing residents. This requires improvements to the built environment, services and job opportunities.

# Questions & Answers

**ⓔ 9/10 marks awarded** The answer to Part (c) scores 4 marks. It covers two different criteria and in both cases these are explained with extended points. The answer goes beyond just stating two criteria to explain how a change in the criteria would show success. The answer to Part (d) scores 5 marks. This is a good answer that shows understanding of re-imaging as part of regeneration. It explains that re-imaging is just one part of the process. It also uses some examples, but these lack depth. Maximum marks could be achieved by adding some place-specific detail for either the Liverpool or Salford examples, perhaps explaining the changes to the built environment or new job opportunities brought by regeneration.

**(e)** Assess the importance of economic restructuring in explaining why some urban areas need to be regenerated.

(12 marks)

ⓔ This extended question uses the command word 'assess' which means 'weigh-up'. The key here is to demonstrate an understanding of economic restructuring, by using examples, but also to consider other explanations of the need for regeneration. These could include demographic change — or even factors such as physical isolation. Using examples to support your answer (evidence) and then making a judgement about the significance of different causes is important. Answers have to focus on urban areas, so avoid using rural examples. Level 3 marks are achieved by making supported judgements about importance. Answers that just explain what economic restructuring is are likely to get no more than 8 marks. Successful arguments would include ones that argue economic restructuring is a crucial element as it then leads to demographic change and environmental decline. Questions in this format do not have a 'right' answer; it is the quality of your assessment and judgements that gain marks. The Levels mark scheme for a 12-mark question is shown below. The same Levels mark scheme is used for the 12-mark question in the 'assess' questions at A-level.

| Level 1<br>1–4 marks | ■ Demonstrates isolated elements of geographical knowledge and understanding, some of which may be inaccurate or irrelevant.<br>■ Applies knowledge and understanding of geographical information/ideas, making limited logical connections/relationships.<br>■ Applies knowledge and understanding of geographical information/ideas to produce an interpretation with limited relevance and/or support.<br>■ Applies knowledge and understanding of geographical information/ideas to make unsupported or generic judgements about the significance of few factors, leading to an argument that is unbalanced or lacks coherence. |
|---|---|
| Level 2<br>5–8 marks | ■ Demonstrates geographical knowledge and understanding, which is mostly relevant and may include some inaccuracies.<br>■ Applies knowledge and understanding of geographical information/ideas logically, making some relevant connections/relationships.<br>■ Applies knowledge and understanding of geographical information/ideas to produce a partial but coherent interpretation that is mostly relevant and supported by evidence.<br>■ Applies knowledge and understanding of geographical information/ideas to make judgements about the significance of some factors, to produce an argument that may be unbalanced or partially coherent. |

| Level 3
9–12 marks | ■ Demonstrates accurate and relevant geographical knowledge and understanding throughout.
■ Applies knowledge and understanding of geographical information/ideas logically, making relevant connections/relationships.
■ Applies knowledge and understanding of geographical information/ideas to produce a full and coherent interpretation that is relevant and supported by evidence.
■ Applies knowledge and understanding of geographical information/ideas to make supported judgements about the significance of factors throughout the response, leading to a balanced and coherent argument. |
|---|---|

## Student answer

(e) Economic restructuring is a key process that causes a need for regeneration. The economic structure of an area refers to its balance of **a** primary, secondary and tertiary employment.

In many UK and USA cities restructuring has meant deindustrialisation and a loss of jobs in manufacturing and distribution. In the case of Middlesbrough **b** this meant a loss of jobs in steel making and petrochemicals, whereas in Detroit **b** the decline of car manufacturing has caused mass unemployment among middle-age male workers in particular. This leads to a spiral of decline **c** or negative multiplier effect whereby the loss of jobs in one sector leads to losses in others — like supplier factories and local services. However, the consequences of restructuring are perhaps more significant, especially population decline. Detroit's population **d** has declined from 1.5 million in 1970 to 0.7 million by 2015. People have left the city because of the loss of jobs, and the lack of new service sector jobs to replace those lost. This is significant because the people who leave tend to be the young and skilled, leaving older less-skilled people behind.

Regeneration is made less likely in cities that gain a reputation **d** for industrial decline and other urban problems like homelessness and crime. This is the case in Liverpool and Detroit and the negative images of these places then deter investment. This shows that although restructuring is a root cause of the need to regenerate it is made worse by demographic and image changes. The shift towards a service sector economy **e** in the USA and UK since 1970 has left many northern industrial cities isolated from the core area of economic growth, i.e. London, and the USA West Coast. This isolation makes it hard to attract investment which tends to go to these core areas.

**e** **12/12 marks awarded** This is a Level 3 answer scoring full marks. It begins with a clear definition **a** of economic structure and restructuring showing an understanding of economic sectors. Examples **b** are used to illustrate restructuring and the use of a concept **c**, the negative multiplier effect, shows deeper understanding and provides an explanation of why some cities need to regenerate. Restructuring is then linked to further changes in population **d** and image **d**, which shows that changes in employment are not the only factor involved. This is an assessment, as the answer begins to consider other factors. The last point about some cities being isolated from the economic core **e** in countries now dominated by the service sector is a useful one, and a further factor to be considered.

# Questions & Answers

## Shaping places: Diverse places

### Question 3

**(a)** Define the term 'ethnicity'. (1 mark)

**(b) (i)** Describe the distribution of areas with a high Asian population shown on Figure 3. (2 marks)

Population %

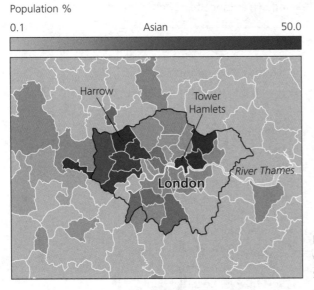

**Figure 3** The Asian population of London in 2011

Source: ONS/economist.com

**(ii)** Suggest one reason why the Asian population is concentrated in some areas of London. (3 marks)

ℯ Part (a) is a recall question requiring a definition. Part (b) requires some analysis of Figure 3 and two clear points about the distribution of the Asian population in London are needed. Remember that some question sequences could contain some marks that test numerical skills. Part (b)(ii) should be answered with reference to one reason only. This means writing a sequence of linked, extended points to gain the 3 marks available.

---

**Student answer**

**(a)** Ethnicity means groups of people that share a common culture, history and traditions as well as language.

**(b) (i)** The Asian population is most heavily concentrated north of the river in London, especially in the northeast and northwest, where several areas have close to 50% Asian population.

**(ii)** The map shows an example of ethnic segregation. This can happen because of internal reasons, such as the desire of an ethnic group to live together for culture and languages reasons, and be close to services such as specific shops and religious buildings. It may also create a feeling of safety, as people are living close to people who share common values.

**ⓔ 6/6 marks awarded** The answer to Part (a) is a correct definition that gains 1 mark. The answer to Part (b)(i) scores 2 marks because there are two clear descriptive points linked to the map shown. The answer to Part (b)(ii) scores 3 marks. This is a good answer that takes the explanation of internal reasons for ethnic segregation and then extends it into explanations about services and safety. An answer about external reasons — which are often negative — would work equally well. Beginning a 3 mark 'explain' question with a broad reason is better than starting with a very narrow one (such as prejudice) because it is hard to extend a very narrow idea.

**(c)** Explain why population density varies across the rural–urban continuum.  (4 marks)

**(d)** Explain the challenges facing rural areas in the UK that have experienced an increase in international immigration.  (6 marks)

ⓔ Part (c) requires an explanation, so there are no marks for describing how population density changes, only for explaining why it changes. Try to avoid trying to cover the whole rural–urban continuum because the question is only worth 4 marks. Part (d) is Levels marked (see Levels mark scheme on page 76), which means examples should be used and you need to cover more than one 'challenge'. Notice this question is specifically about rural areas, so do not mention large towns and cities as this will score no marks. A small market town in a rural area would be acceptable. Equally, answers need to focus on international not internal migration.

### Student answer

**(c)** High population densities are found in cities because people want to be close to jobs, so houses are small and tall, increasing density especially in inner cities. In suburbs, densities are lower because people demand large homes with gardens and can afford to pay for this. In rural areas, few workers are needed on farms so settlement is sparse and population density low, especially in the most physically inaccessible locations which have little economic activity.

**(d)** Some rural areas, such as Lincolnshire, Cambridgeshire and especially the market town of Boston in Lincolnshire have seen a rapid rise in EU A8 migrants since 2004. In Boston they account for about 16% of the population. Population increases of 10% plus in rural areas are large, and this puts pressures on housing supply and may increase prices and rents, making it harder for young people to find a home. Schools face challenges of educating migrant children and may need translators, which adds to school costs. Significantly in small rural schools even an extra 5–10 pupils could represent a large increase. There have been some cultural tensions in Boston caused by changes to services, e.g. Polish shops and a feeling of rapid cultural change; however, these are relatively small scale. Nevertheless support for nationalist political parties such as UKIP is especially strong in Lincolnshire, suggesting the host population feels uncomfortable with the pace of immigration.

🅔 **10/10 marks awarded** The answer to Part (c) scores 4 marks. It covers a number of different locations across the continuum including inner cities, suburbs and inaccessible rural areas and shows good understanding of the concept of the rural–urban continuum. There are explanations based on opportunities in cities, incomes and limited economic activity in rural areas.

The answer to Part (d) scores 6 marks. This is a located answer using Lincolnshire and Boston to illustrate some issues in rural areas. The data provided add depth and a range of challenges are explained including housing, education and cultural issues. This range shows understanding, as do the comments relating to the significance of some challenges compared to others. As always with questions about immigration, answers need to be considered, balanced and based on evidence.

**(e)** **Assess the roles of different stakeholders in managing change in demographically and culturally diverse urban areas.** (12 marks)

🅔 This extended question uses the command word 'assess' which means 'weigh-up'. The key here is to demonstrate understanding of a range of situations where different stakeholders have been involved in managing population change and cultural change. Make sure your answer covers a range of different stakeholders — three would be a good rule of thumb. What you do not want to write is just an extended list of stakeholders. The 'assess' element can be considered by thinking about which stakeholders are the most important in terms of managing change — which make the biggest difference or have the most influence? Level 3 marks are achieved by making supported judgements about the roles and significance of stakeholders (see Levels mark scheme on pages 80–81). Answers that just explain different roles are likely to get no more than 8 marks. Successful arguments would include ones that argue that government or local councils are the most significant, or argue that demographic change is more successfully managed than cultural (or vice versa). Questions in this format do not have a 'right' answer; it is the quality of your assessment and judgements that gains marks.

### Student answer

**(e)** The UK has experienced cultural and demographic change especially since the year 2000 as immigration numbers have risen, partly as a result of A8 migration since 2004 and an increase in birth rate. This means population 🅐 is expanding faster than for several decades, although it is regionally unbalanced with the South and East growing most rapidly. A key stakeholder is national government. It manages immigration volume, and has introduced a tiered system for non-EU immigrants to try to reduce numbers. However, 🅓 it currently cannot manage EU immigration due to the open borders EU policy 🅑. In that sense the EU is a more significant player than national government.

Rising population leads to a demand for housing which is managed at a local level by councils 🅒. Many councils have struggled to meet housing needs because of a shortage of available land, partly due to planning restrictions such as national government greenbelt policy. In Southwark housing shortages have been met by a policy of widespread regeneration of council housing estates such as Aylesbury and Heygate **b** which use private developers and housing associations to regenerate run-down council housing at increased densities. Local community groups such as 'Just Space' **c** criticise this approach, arguing that it leads to rising rents and prices that low income groups including ethnic minority groups cannot afford. Luton has experienced a cultural change with a rise in population of the Asian ethnic minority group since the 1960s. Asians are unusually heavily segregated in Luton, especially concentrated in the area of Bury Park **b**.

Cultural tensions in Luton between the Muslim population, some existing residents and nationalist political groups have grown in recent years. These issues should **d** have been managed by the local council whose housing policies could have helped limit segregation. In addition, community groups on both sides have a key role in managing change to ensure mutual understanding and minimising conflict. Overall **e**, it is perhaps local councils like Luton and Southwark that are the key stakeholders as cultural and demographic change are local issues because some places are affected by them much more than others.

**e** **12/12 marks awarded** This is a Level 3 answer scoring full marks. It deals with both demographic and cultural change **a**, which is a requirement to gain Level 3 marks. The answer is supported by examples including the EU, Luton and Southwark **b** and there is some detail about these places and stakeholders — such as the reference to specific housing estates (Aylesbury, Heygate) and small areas (Bury Park). A range of stakeholders **c** is also considered including the EU, national and local councils and community groups. The answer also contains some assessment **d** including the idea that the EU (as things stood in 2016) was more significant than national government in terms of managing migration and that local councils are perhaps the key stakeholder in managing change. The last point about change being essentially local **e** is very true.

# ■A-level questions

## Globalisation

## Question 1

(a) Study Table 1.

| World's 500 largest TNCs in 2016 by country of origin | | World's 10 largest TNCs in 2016 by revenue | |
|---|---|---|---|
| Country | Number of TNCs | TNC name | Revenue (US $ billions) |
| USA | | Walmart | 482.1 |
| China | 103 | State Grid | 329.6 |
| Japan | 52 | CNP | 299.3 |
| France | 29 | Sinopec | 294.3 |
| Germany | 28 | Royal Dutch Shell | 272.2 |
| UK | 26 | Exxon Mobil | 246.2 |
| South Korea | 15 | VW | 236.6 |
| Switzerland | 15 | Toyota | 236.6 |
| Netherlands | 12 | Apple | 233.7 |
| Canada | 11 | BP | 225.8 |
| Other countries | 75 | | |
| **TOTAL** | **500** | **TOTAL** | **2856.4** |

**Table 1** Data on the world's largest TNCs in 2016

Calculate the following using the data in Table 1. (4 marks)

(i)   The number of TNCs that originate in the USA.

(ii)  The percentage (%) of TNCs that originate from China.

(iii) The mean revenue of the 10 largest TNCs.

(iv) The difference in revenue between Walmart and BP.

**e** Part (a) is a skills question, using the information in Table 1. It requires some careful maths (you are allowed to use a calculator) and should not be rushed. The calculations are not demanding but you do need to be accurate. Calculate your answers to one decimal place.

---

**Student answer**

(a) (i)   134

    (ii)  20.6%

    (iii) $285.6

    (iv) $256.3

---

(e) **4/4 marks awarded** The calculations are all correct so the answer scores full marks. The answer to (a)(i) is calculated by adding up the values in column two and subtracting this value from 500. The answer to (a)(ii) is a simple percentage calculation 103 / 500 × 100 = 20.6%. The answer to (a)(iii) simply involves shifting the decimal place in the total revenue cell because there are ten TNCs (alternatively add up the ten revenues and divide by 10). The answer to (a)(iv) involves subtracting 225.8 from 482.1 to find the difference.

**(b)** Assess the extent to which local groups and NGOs can offset some of the negative consequences of globalisation. (12 marks)

(e) Part 1(b) is an extended writing question that is marked in Levels. 'Assess the extent' means 'how far' so your answer needs to be evaluative in style. Both local groups and NGOs need to be mentioned, and high quality answers will also consider other groups and players that perhaps cause some of the negative consequences of globalisation. Is the work of local groups and NGOs enough to offset these consequences? Rather than thinking about negative consequences as being one thing, it needs to be split into social, economic and environmental aspects. Examples need to be used to support your answer to give it the required depth. The Levels mark scheme for 12 mark questions is shown below. The same Levels mark scheme is used for the 12 mark questions in the AS options questions, and 12 mark 'assess' questions at A level.

| Level 1<br>1–4 marks | ■ Demonstrates isolated elements of geographical knowledge and understanding, some of which may be inaccurate or irrelevant.<br>■ Applies knowledge and understanding of geographical information/ideas, making limited logical connections/relationships.<br>■ Applies knowledge and understanding of geographical information/ideas to produce an interpretation with limited relevance and/or support.<br>■ Applies knowledge and understanding of geographical information/ideas to make unsupported or generic judgements about the significance of few factors, leading to an argument that is unbalanced or lacks coherence. |
|---|---|
| Level 2<br>5–8 marks | ■ Demonstrates geographical knowledge and understanding, which is mostly relevant and may include some inaccuracies.<br>■ Applies knowledge and understanding of geographical information/ideas logically, making some relevant connections/relationships.<br>■ Applies knowledge and understanding of geographical information/ideas to produce a partial but coherent interpretation that is mostly relevant and supported by evidence.<br>■ Applies knowledge and understanding of geographical information/ideas to make judgements about the significance of some factors, to produce an argument that may be unbalanced or partially coherent. |
| Level 3<br>9–12 marks | ■ Demonstrates accurate and relevant geographical knowledge and understanding throughout.<br>■ Applies knowledge and understanding of geographical information/ideas logically, making relevant connections/relationships.<br>■ Applies knowledge and understanding of geographical information/ideas to produce a full and coherent interpretation that is relevant and supported by evidence.<br>■ Applies knowledge and understanding of geographical information/ideas to make supported judgements about the significance of factors throughout the response, leading to a balanced and coherent argument. |

# Questions & Answers

(b) One of the most widely reported negative consequences of globalisation is the ill-treatment of workers in emerging countries who work in sweatshop conditions for ⏹c low pay and in very poor working conditions. Despite this being well known, consumers continue to buy products like jeans made by low paid textile workers in Bangladesh. Few consumers boycott these products or are prepared to seek out alternatives. An exception is food products like tea, coffee and cocoa which are part of the ⏹a NGO Fairtrade Foundation. Fairtrade pays developing world farmers a higher price for their produce, so farmers can enjoy a better quality of life. Consumers pay more for the Fairtrade version of the product. Fair trade does work, but only for a small number of farmers so has ⏹d limited impact. Most coffee and cocoa traded globally is not fair trade. Critics of fair trade point out that the extra income received by farmers is small anyway.

Environmentally, the ⏹a Forest Stewardship Council NGO tries to limit global deforestation by labelling products as coming from sustainably managed forests. Like the Fairtrade logo, the FSC logo is well known and has raised global awareness. In both cases, this is perhaps the ⏹d main success of these NGOs in that they have made consumers aware of globalisation's ⏹c social, economic and environmental down sides, which can slowly begin to change consumer behaviour. Some well-known brands, like ⏹b Cadbury, are now Fairtrade showing it is now mainstream.

⏹a Transition towns are a local attempt to promote locally grown and made produce and encourage consumers to buy and trade locally, thus protecting local jobs from the march of global products and brands. ⏹b Totnes is an example, encouraging people to grow their own food to reduce the ⏹c environmental impact of food transport. These local schemes are small, with a ⏹d limited impact. They perhaps serve to make a small group of people feel good, rather than actually ⏹d denting the low pay and poor working conditions suffered by many food producers and factory workers in emerging and developing countries.

ⓔ **12/12 marks awarded** The answer to (b) scores full marks. It covers both NGOs and local groups ⏹a and is therefore balanced in terms of the demands of the question. Examples ⏹b are used and there is some detail about some of the groups mentioned. Rather than just being about 'consequences' in general there is some attempt to identify ⏹c environmental consequences and socio-economic consequences, which helps show depth of understanding. There are also some ⏹d judgements, i.e. assessment of how far the schemes mentioned have made a difference. It is important to state that the judgements are quite realistic, as in reality NGOs and local schemes have a limited (but positive) impact.

# Shaping places: Regenerating places

## Question 2

**(a)** Study Figure 4. Explain the relationship between level of deprivation and percentage of people living close to derelict land. (6 marks)

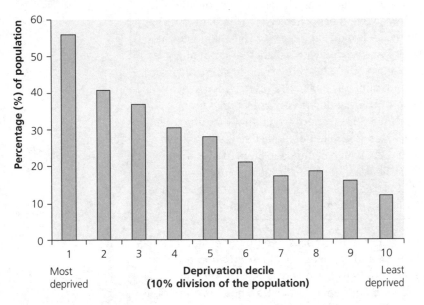

**Figure 4** Percentage of Scotland's population living within 500 metres of derelict land by deprivation decile in 2014
Source: www.gov.scot

**e** This is a data stimulus question. The resource, in this case a bar chart, needs to be studied carefully and evidence from it used to answer the question. Basically it shows that the most deprived are more likely to live close to derelict land. As deprivation levels fall, the percentage of people living close to derelict land falls. The differences are quite large, ranging from 55% to 15%. Notice the large drop between decile 1 and 2, and the fact that decile 8 appears to be an anomaly — although not a large one: it could be accounted for by variation in the data set. The key is to provide two or three convincing, detailed reasons for this relationship. These 6-mark questions in the Shaping places topic are marked using the following Levels mark scheme.

| Level 1<br>1–2 marks | ▪ Demonstrates isolated or generic elements of geographical knowledge and understanding, some of which may be inaccurate or irrelevant.<br>▪ Applies knowledge and understanding to geographical information inconsistently. Connections/relationships between stimulus material and the question may be irrelevant. |
|---|---|
| Level 2<br>3–4 marks | ▪ Demonstrates geographical knowledge and understanding which is mostly relevant and may include some inaccuracies.<br>▪ Applies knowledge and understanding to geographical information to find some relevant connections/relationships between stimulus material and the question. |
| Level 3<br>5–6 marks | ▪ Demonstrates accurate and relevant geographical knowledge and understanding throughout.<br>▪ Applies knowledge and understanding to geographical information logically to find fully relevant connections/relationships between stimulus material and the question. |

**Student answer**

**(a)** Figure 4 shows that as deprivation increases people are more likely to live close to derelict land. This is because derelict land is linked to deindustrialisation and loss of jobs such as closed factories, which lowers income and leads to deprivation. Derelict sites are unattractive and could lower house prices, as well as deterring new development, so low income deprived groups are more likely to live there. There is a large drop of 15% between the most deprived groups 1 and 2 on Figure 4. This suggests the highest levels of deprivation are associated with derelict land but slightly less deprived areas are much less so. The least deprived people can afford to move away from derelict land, and may live in the suburbs or rural areas where there is less derelict land. Deprivation is concentrated in inner cities and so is dereliction because this is where traditional manufacturing industry once was. Decile 8 is an anomaly, but actually deciles 7–10 have very similar percentages of 12–20% which could be explained by natural variation in the data. People in the least deprived areas may pressure councils to clear up derelict land because they are more politically engaged and economically powerful and can influence decision making.

**e** **6/6 marks awarded** This answer scored full marks. It both understands the general relationship shown in Figure 4 and explains the apparent anomaly in a sensible way. There is a range of explanations including deindustrialisation as a cause of dereliction which then leads to deprivation as well as good explanations relating to house prices and the fact that higher income groups can afford to choose where to live. The last point about political engagement is especially good. This type of data driven question often does not lend itself to the use of examples: it's more about providing clear explanations and referring to the data provided.

**(b)** Explain how rural areas can be regenerated using contrasting strategies. (6 marks)

**e** This style of question is best thought of as a 'mini-essay'. The topic is quite a narrow one focused on rural regeneration only. There is no stimulus material so you need to use detailed knowledge and understanding to provide explanations. Examples, but not large case studies, can also be used to support your explanations. In this question answers need to explain contrasting strategies so choose a small-scale example such as farm diversification plus a larger scheme such as the Eden Project or Trump International Golf Links. These 6 mark questions in the Shaping places options are marked using the following Levels mark scheme.

| Level 1<br>1–2 marks | ■ Demonstrates isolated or generic elements of geographical knowledge and understanding, some of which may be inaccurate or irrelevant.<br>■ Applies knowledge and understanding to geographical information inconsistently. Connections/relationships between stimulus material and the question may be irrelevant. |
|---|---|
| Level 2<br>3–4 marks | ■ Demonstrates geographical knowledge and understanding, which is mostly relevant and may include some inaccuracies.<br>■ Applies knowledge and understanding to geographical information to find some relevant connections/relationships between stimulus material and the question. |
| Level 3<br>5–6 marks | ■ Demonstrates accurate and relevant geographical knowledge and understanding throughout.<br>■ Applies knowledge and understanding to geographical information logically to find fully relevant connections/relationships between stimulus material and the question. |

**Student answer**

**(b)** Rural areas experience problems linked to the post-production countryside and the need to find jobs to replace those lost in farming or mining. Closed mines and abandoned farms can create derelict rural land so economic and environmental regeneration is needed. Much rural regeneration is on the farm-scale, with individual farms and estates using diversification as a way of creating new incomes and jobs, and re-purposing unused buildings. The Milkhope Centre near Newcastle is an example, where unused farm buildings have become retail outlets, workshops and cafes. Farmers usually manage the process themselves, although EU grants are often available.

On a different scale rural regions are often regenerated by re-imaging with a 'brand' often led and funded by one or more local council. Hardy Country in Dorset is an example. Dorchester and surrounding areas are advertised as being linked to the author Thomas Hardy. The aim here is to raise tourism awareness, attract people to the area and increase tourism and leisure spending. In some cases derelict land can be directly repurposed. Examples include the Eden Project in Cornwall (a garden and park with exotic 'biome' domes) and the Blaenau Ffestiniog abandoned slate quarries (mountain biking, caving and zipwire adventure) and hills in Wales. In both of these cases specific tourist attractions have been created in abandoned quarries that act as a focus for tourist visits and might create demand for accommodation and other tourist services.

Rural regeneration on a large scale can attract opposition in areas not used to change. The Trump International Golf Links, a potential £1 billion private investment funded coastal golf resort on the Menie Estate in Balmedie, Aberdeenshire has faced major local opposition despite promises of jobs, investment and a boosted local economy. Environmental and landscape issues as well as concerns over the scale of development often mean rural regeneration is small, and projects like the Eden Project are an exception.

**e** **6/6 marks awarded** This is a Level 3 answer that scored full marks. It uses good terminology that shows an understanding of regeneration in a rural context. There are lots of examples, all of which are clearly rural, and which are applied to the question. A key issue is whether the examples 'contrast' as that is the focus of the question. The answer is that they do: the candidate starts with very small-scale, farmer-led, diversification and moves on to regional strategies that are really about marketing to promote an area and support regeneration. Finally several large-scale schemes (at least in terms of rural regeneration) are considered. There is also some contrast provided by type of funding source, as well as an attempt to recognise that not all strategies are successful. This is not required by the command word 'explain' but it does help to show an in-depth understanding of rural regeneration strategies.

**(c)** Evaluate the extent to which urban regeneration usually meets the needs of all stakeholders involved in the regeneration process. (20 marks)

ⓔ This question is an essay question with a high mark tariff. The question has a number of different elements to it (urban + stakeholders + regeneration + needs), all of which need to be covered. There are concepts that need to be addressed too, i.e. what is regeneration and what might its 'needs' be— these will vary by stakeholder as the needs of investors might be profit, whereas existing residents' needs might be better services and opportunities. The command word 'evaluate' means 'weigh-up and come to a judgement'. Good answers will argue a case using examples and case studies to back up the argument. In the case of this question it might be argued that it is very rare that the needs of all stakeholders are met, because their needs are so different. Examples that are mostly successful versus ones that clearly have failed to meet some stakeholders' needs are useful — the contrast allows you to get an argument going and make a judgement about success. These 20 mark questions in the Shaping places options are marked using the following Levels mark scheme.

| | |
|---|---|
| **Level 1**<br>**1–5 marks** | ■ Demonstrates isolated elements of geographical knowledge and understanding, some of which may be inaccurate or irrelevant.<br>■ Applies knowledge and understanding of geographical ideas making limited and rarely logical connections/relationships.<br>■ Applies knowledge and understanding of geographical information/ideas to produce an interpretation with limited coherence and support from evidence.<br>■ Applies knowledge and understanding of geographical information/ideas to produce an unsupported or generic conclusion, drawn from an argument that is unbalanced or lacks coherence. |
| **Level 2**<br>**6–10 marks** | ■ Demonstrates geographical knowledge and understanding, which is occasionally relevant and may include some inaccuracies.<br>■ Applies knowledge and understanding of geographical information/ideas with limited but logical connections/relationships.<br>■ Applies knowledge and understanding of geographical ideas in order to produce a partial interpretation that is supported by some evidence but has limited coherence.<br>■ Applies knowledge and understanding of geographical information/ideas to come to a conclusion, partially supported by an unbalanced argument with limited coherence. |
| **Level 3**<br>**11–15 marks** | ■ Demonstrates geographical knowledge and understanding, which is mostly relevant and accurate.<br>■ Applies knowledge and understanding of geographical information/ideas to find some logical and relevant connections/relationships.<br>■ Applies knowledge and understanding of geographical ideas in order to produce a partial but coherent interpretation that is supported by some evidence.<br>■ Applies knowledge and understanding of geographical information/ideas to come to a conclusion, largely supported by an argument, that may be unbalanced or partially coherent. |
| **Level 4**<br>**16–20 marks** | ■ Demonstrates accurate and relevant geographical knowledge and understanding throughout.<br>■ Applies knowledge and understanding of geographical information/ideas to find fully logical and relevant connections/relationships.<br>■ Applies knowledge and understanding of geographical information/ideas to produce a full and coherent interpretation that is supported by evidence.<br>■ Applies knowledge and understanding of geographical information/ideas to come to a rational, substantiated conclusion, fully supported by a balanced argument that is drawn together coherently. |

## Student answer

**(c)** Urban regeneration **a** involves renewing the built environment of a place as well as regenerating services like health and education in order to attract inward investment and improve quality of life. It can be a very controversial process and in some cases some stakeholders feel their needs are sidelined **b**. Major regeneration schemes often seek to find a new purpose for large areas of unused and derelict former industrial land. This was the case in the 1980s in **c** London's Docklands and in the 2000s in **c** Salford Quays. In both cases, existing residents felt their need for better housing, services and suitable jobs, i.e. relatively low skills in these deprived areas, were not prioritised because regeneration focused on private housing for sale, office developments and flagship schemes such as Salford's MediaCity where the **d** BBC employed about 30 local people out of a workforce of 700. Large schemes are either run by **e** quangos, or private investors such as **d** Peel Holdings in the case of Salford Quays. They are motivated mainly by profit, and this seems to limit the benefits that come to local people. Even the **c** 2012 London Olympic regeneration in Stratford saw protests by residents of the **d** Clay Lane Housing Co-operative, which was cleared to make way for Olympic Games infrastructure.

**f** On the other hand services, access via tube and bus and housing have all improved in Stratford. Many of the athletes' village apartments have since become **e** affordable housing and the new Westfield Shopping Centre and legacy sports faculties do employ local people. Many councils in England need to regenerate **e** 'sink estates', many of which date back to the 1960s and are now run-down pockets of very high deprivation. **d** Southwark Council in London is regenerating the **c** Heygate and Aylesbury estates. Because regeneration costs are so high a combination of housing associations and private developers are leading the regeneration. This has led to accusations of spiralling 'affordable rents' and **e** gentrification as private developers build luxury apartments on some of the land sold to them by Southwark Council. Local community groups such as **d** Just Space and 35percent have emerged to fight this style of regeneration. **f** In this example, housing is improved for many people, of course.

**f** Overall, regeneration often meets the needs of many stakeholders. If the image of an area is improved local businesses and many residents will benefit as investment and services are attracted back to the area. Private investors such as house builders usually profit, and councils have fewer environmental and crime issues to deal with in improved areas. However **f**, because regeneration usually involves large scale change to the built environment there is often a minority of local people who feel they are excluded and often these are the poorest and least skilled.

**ⓔ 20/20 marks awarded** This answer is Level 4 and scores full marks. It starts with a definition of regeneration **ⓐ** which shows good understanding of the process. There is also an early statement of what the argument is going to be **ⓑ**, which is a useful way of providing some clarity. The answer uses a range of relevant examples **ⓒ** with some details on location and **ⓓ** specific stakeholders. This approach is better than trying to rely on one case study, which tends to produce a descriptive answer. There is good use of **ⓔ** terminology throughout, which shows the topic is understood in detail. Evaluation **ⓕ** is present as well. This includes the argument that despite some local opposition the Olympic site has generally benefited the area of Stratford, and the overall conclusion that a small number of local people are often excluded from regeneration and feel they do not benefit.

# Shaping places: Diverse places

## Question 3

**(a)** Study Figure 5. Explain the changes to the population of England and Wales shown in Figure 5.

(6 marks)

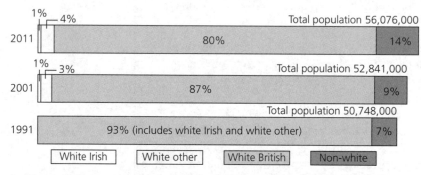

**Figure 5** Changes to the population of England and Wales 1991–2011

**ⓔ** This is a data stimulus question. The resource, in this case a bar chart, needs to be studied carefully and evidence from it used to answer the question. The bar chart shows both changing ethnicity and changing total population — both elements need to be explained with reference to both immigration and internal growth (birth rate): it is important to recognise that the changes are not only a result of immigration. The immigration element can be related to both EU and non-EU movements.

---

**Student answer**

**(a)** The first change is the increase in population of 2 million between 1991 and 2001 and then 3.5 million up to 2011. This can be explained by rising fertility and higher birth rates causing population growth. The accelerating population growth is a result of higher levels of immigration, especially since 2004 when free movement of workers from the A8 countries started. Higher levels of immigration further boosted total population above birth rate increase. In addition, immigrants tend to be young and have higher fertility, so they have in turn increased birth rate. Secondly the increase in non-white population is partly due to immigration but also the existing UK black and Asian population having children so increasing the percentage.

The increase in white-other percentage, and white-Irish, has been caused by the freedom of movement of EU citizens coming to the UK to work. Many Poles and other eastern Europeans, plus people from France and Spain, have migrated to the UK because of the relative strength of the UK economy.

**e** **6/6 marks awarded** This answer scored full marks (see Levels mark scheme on page 89). It is a logical answer that begins with total numbers and progresses to explain the change in ethnicity. It also recognises the more complex fact that immigration and birth rate are linked; immigration has caused natural increase so the two can't be seen as separate explanations. There is a danger with Figure 5 that answers only focus on immigration, but this is not the only explanation for the population changes shown. The answer has a good understanding of how EU migration relates to the data shown and uses good terminology. It is an answer that analyses Figure 5, then explains it.

**(b)** Explain the processes that lead to ethnic clustering in some locations. (6 marks)

**e** This style of question is best thought of as a 'mini-essay'. The topic is quite a narrow one focused on the concept of ethnic clustering (segregation). There is no stimulus material so you need to use detailed knowledge and understanding to provide explanations. Examples, but not large case studies, can also be used to support your explanations. With this topic, it is important to give a balanced explanation of the internal and external causes of clustering, rather than provide a very one-sided explanation, e.g. one that focuses only on prejudice and racism (external causes).

### Student answer

**(b)** In cities, ethnic clustering is common. It is segregation based on the ethnicity of a community rather than clustering based on income levels alone. In Luton for example the Asian ethnic minority is tightly clustered northwest of the CBD including a majority Muslim area called Bury Park. In the London borough of Southwark there are clusters of British Asian and black British groups. In many cases these areas are low income, but not always. There are middle-class Asian clusters in Harrow (London) and Oadby in Leicester. This provides a clue that low incomes and high deprivation alone cannot explain segregation. It is often explained with reference to internal and external factors, which are largely but not entirely positive and negative respectively. Internal factors include the tendency of ethnic minorities to cluster because they share a common language and may not speak the host population language.

In addition clustering allows communities to easily share religious buildings, specific shops and schools that would otherwise be hard to access. Early immigrants may create a community, like Bangladeshi Brick Lane in east London, which then becomes an attractive area for later immigrants. Ethnic minority groups may cluster for mutual protection, or worse, fear of the host society.

> External factor explanations focus on the attitudes and actions of the host. These may include prejudice and discrimination in the jobs and housing market, which means ethnic minority groups have lower incomes and therefore less choice of where to live. Estate agents and council housing managers may effectively exclude ethnic groups from some places. The host community can react by moving out of an area that an ethnic minority moves into, thus creating segregation rather than accepting mixing.

**ⓔ 6/6 marks awarded** This is a Level 3 answer scoring full marks (see Levels mark scheme on page 90). This is a good answer. It begins with some examples of ethnic clustering which demonstrate a good understanding of segregation and where it is found. This part of the answer does not provide explanations but it does provide context, which the explanations provided later on then link to. It also shows the candidate understands that ethnic segregation is about more than poverty — through the use of the 'middle-class' examples.

The majority of the answer is then structured by internal and external causes which make logical sense. Both broad explanations are broken down into different sub-elements such as the housing market and benefits of clustering close to services. Overall, this is a full and balanced answer that shows good understanding of ethnic clustering and uses examples to support this.

**(c)** Evaluate the economic and social significance of demographic change in the UK in recent decades for contrasting places. (20 marks)

**ⓔ** This is an essay question with a high mark tariff. The question has a number of different elements to it (economic + social + demographic + contrasting) all of which need to be covered. The command word 'evaluate' means 'weigh-up and come to a judgement'. This is the sort of question that needs some thinking and planning time. Which examples could be used that are contrasting? Anglesey and Southwark would work well here as they have a rural/urban, decline/growth and ethnicity and age contrast. Good answers will argue a case using examples and case studies to back up the argument.

For this question, places like Anglesey have seen population decline and the loss of young people as well as economic decline leaving them isolated and lacking diversity. Places like Southwark have seen dramatic changes in terms of ethnicity and population expansion putting pressure on the area but creating economic success as well as deprivation. Anglesey's demographic changes could be judged as resulting in the place becoming more like it already was, i.e. rural, isolated and ageing, whereas in Southwark the demographic changes might be viewed as more transformative.

## Student answer

**(c)** Many places in the UK have undergone change in terms of **a** population numbers, structure and ethnicity in the last 40 years but the significance of this varies. One approach is to examine the changes at different **b** scales from national to local in order to evaluate their economic and social significance.

Nationally **c** the total population of the UK has 'tipped' towards the South. Between 1981 and 2011 there was barely any change in the population of Scotland, the North East and North West but more than 1 million new people in each of London, the South and South West. The economic significance of the South and London has grown due to booming **e** service industries, whereas declining secondary industry has weakened the North and also led to internal migration to the South.

Two small areas can be used to illustrate the causes and consequences. Anglesey **c** in northwest Wales has a declining, ageing and ethnically white population whereas the borough of **c** Southwark in London has a youthful, rapidly rising and ethnically diverse population. Industrial and farming decline on Anglesey has pushed out its **d** young people, whereas rapid economic growth in London has attracted internal and especially international migrants to Southwark — its population was 46% from ethnic minority groups in 2011 compared to 2% in Anglesey. Socially, Anglesey is dominated by an **d** ageing dependent population with a lack of young people, which limits its **e** economic prosperity. Southwark is dominated by 20–40-year-old working-age people. Many of these people have good incomes **f**, but rapidly rising population has pushed up house prices so **d** living costs are high and disposable **e** income is often low.

At a very small scale Southwark's wards illustrate another demographic change. Rising international immigration has led to an ethnically diverse Southwark, but ethnic minorities tend to be clustered in particular locations. Examples include the black African population in **c** Peckham and Livesey wards and Bangladeshis in **c** Walworth. Socially, this segregation is **f** significant because it is often associated with deprivation. Ethnic minority groups on average have **e** lower incomes and **d** poorer housing. Many **f** UK inner cities have pockets of this type of deprivation and this can lead to social tensions as ethnic minority groups feel marginalised and ignored. Interestingly, remote rural locations such as Anglesey may feel equally **d** marginalised, although for different reasons, i.e. being increasingly distant from the **e** economic core of the UK in London and the South.

Overall **g**, demographic change has increased the economic contrast between a successful, urban South and a less prosperous North (both urban and rural) as well as increasing cultural diversity in the South more than in the North. However locally, increasing ethnic clustering has created a **g** significant problem of socially and economically deprived inner city locations.

**ⓔ 20/20 marks awarded** This is a Level 4 answer that scored full marks (see Levels mark scheme on page 92). This is a very good answer. At the start the answer breaks down the term **ⓐ** 'demographic' into its component parts, i.e. numbers, structure and ethnicity, which shows good understanding. There is a very useful attempt to organise the answer by **ⓑ** scale, which shows the candidate has recognised the importance of a logical structure. A risk with this sort of question is that it is answered with reference to numerous places and becomes very random, or one big case study is used which would lead to a narrow answer. This answer chooses a middle path so there are enough **ⓒ** examples to provide depth but few enough to maintain some order. There are also both **ⓓ** social and **ⓔ** economic explanations linking demographic change to social and economic change. In some cases these overlap, but this is acceptable as social and economic issues are intertwined. The **ⓕ** significance of the changes is also made clear, which is the 'evaluation' element of the question. There is also a useful conclusion which acts as a final evaluation **ⓖ** arguing that local inner-city changes may be the most significant.

## Knowledge check answers

1 Ships, lorries/trucks and trains.
2 Subsea fibre optic cables.
3 The WTO (World Trade Organisation).
4 1978.
5 Wal-mart Stores.
6 1950.
7 About 50%.
8 Lagos.
9 Money sent home by migrants.
10 Obesity, diabetes (also heart disease).
11 HDI or GII.
12 Income inequality / distribution.
13 1800.
14 The emigrant population of a country living abroad.
15 Wood products are from sustainable forests.
16 The secondary sector.
17 Four years.
18 It declined, by 7600 people.
19 There are seven domains.
20 Two from Verizon, Oracle, Microsoft, Cisco, PepsiCo and Vodafone.
21 Black in Detroit, white in Santa Clara County.
22 Young people in their late teens and twenties.
23 General elections.
24 A sink estate.
25 The Census, the IMD, labour force surveys.
26 Local areas, via LEPs.
27 £470,000.

28 Quaternary sector/hi-tech businesses and jobs.
29 The history of an area and its historic sites.
30 The Beatles.
31 Studentification.
32 Their health.
33 It is often contaminated.
34 Shipping/the dock trade.
35 Tertiary (leisure and tourism are part of the service sector).
36 London, the South East and the South West.
37 Remote rural areas.
38 The 20–40 age group.
39 Black African.
40 They are Welsh speakers.
41 The suburbs.
42 The suburbs.
43 Heating/energy/transport/food costs.
44 Television programmes.
45 Moving out of an urban area to live in the countryside.
46 London.
47 1948.
48 British Bangladeshis.
49 Gentrification.
50 Clustering or segregation.
51 Bangladeshi.
52 Village.
53 PwC.
54 Tourism.
55 Young people/young families.

# Index

Note: **bold** page numbers indicate defined terms.

# Index

# Index